William
Morgan

T0345250

SCIENTISTS OF WALES

William Morgan

EIGHTEENTH-CENTURY ACTUARY, MATHEMATICIAN AND RADICAL

NICOLA BRUTON BENNETTS

UNIVERSITY OF WALES PRESS

2020

www.uwp.co.uk

British Library Cataloguing-in-Publication Data
A catalogue record for this book is available from the British Library.

ISBN   978-1-78683-618-2
eISBN   978-1-78683-619-9

The right of Nicola Bruton Bennetts to be identified as author of this work has been asserted in accordance with sections 77, 78 and 79 of the Copyright, Designs and Patents Act 1988.

The publisher acknowledges the financial support of the Books Council of Wales.

THE LEARNED SOCIETY OF WALES
CYMDEITHAS DDYSGEDIG CYMRU

MIX
Paper from
responsible sources
FSC   FSC® C013604
www.fsc.org

Typeset by Marie Doherty
Printed by CPI Antony Rowe, Melksham
Reprinted 2021

In memory of my brother, Simon Bruton
(1947–2019)

CONTENTS

SERIES EDITOR'S FOREWORD

Wales has a long and important history of contributions to scientific and technological discovery and innovation stretching from the Middle Ages to the present day. From medieval scholars to contemporary scientists and engineers, Welsh individuals have been at the forefront of efforts to understand and control the world around us. For much of Welsh history, science has played a key role in Welsh culture: bards drew on scientific ideas in their poetry; renaissance gentlemen devoted themselves to natural history; the leaders of early Welsh Methodism filled their hymns with scientific references. During the nineteenth century, scientific societies flourished and Wales was transformed by engineering and technology. In the twentieth century the work of Welsh scientists continued to influence developments in their fields.

Much of this exciting and vibrant Welsh scientific history has now disappeared from historical memory. The aim of the Scientists of Wales series is to resurrect the role of science and technology in Welsh history. Its volumes trace the careers and achievements of Welsh investigators, setting their work within their cultural contexts. They demonstrate how scientists and engineers have contributed to the making of modern Wales as well as showing the ways in which Wales has played a crucial role in the emergence of modern science and engineering.

RHAGAIR GOLYGYDD
Y GYFRES

O'r Oesoedd Canol hyd heddiw, mae gan Gymru hanes hir a phwysig o gyfrannu at ddarganfyddiadau a menter gwyddonol a thechnolegol. O'r ysgolheigion cynharaf i wyddonwyr a pheirianwyr cyfoes, mae Cymry wedi bod yn flaenllaw yn yr ymdrech i ddeall a rheoli'r byd o'n cwmpas. Mae gwyddoniaeth wedi chwarae rôl allweddol o fewn diwylliant Cymreig am ran helaeth o hanes Cymru: arferai'r beirdd llys dynnu ar syniadau gwyddonol yn eu barddoniaeth; roedd gan wŷr y Dadeni ddiddordeb brwd yn y gwyddorau naturiol; ac roedd emynau arweinwyr cynnar Methodistiaeth Gymreig yn llawn cyfeiriadau gwyddonol. Blodeuodd cymdeithasau gwyddonol yn ystod y bedwaredd ganrif ar bymtheg, a thrawsffurfiwyd Cymru gan beirianneg a thechnoleg. Ac, yn ogystal, bu gwyddonwyr Cymreig yn ddylanwadol mewn sawl maes gwyddonol a thechnolegol yn yr ugeinfed ganrif.

Mae llawer o'r hanes gwyddonol Cymreig cyffrous yma wedi hen ddiflannu. Amcan cyfres Gwyddonwyr Cymru yw i danlinellu cyfraniad gwyddoniaeth a thechnoleg yn hanes Cymru, â'i chyfrolau'n olrhain gyrfaoedd a champau gwyddonwyr Cymreig gan osod eu gwaith yn ei gyd-destun diwylliannol. Trwy ddangos sut y cyfrannodd gwyddonwyr a pheirianwyr at greu'r Gymru fodern, dadlennir hefyd sut y mae Cymru wedi chwarae rhan hanfodol yn natblygiad gwyddoniaeth a pheirianneg fodern.

ACKNOWLEDGEMENTS

This book began with letters from the past which introduced me to distant members of my family. I am forever grateful to John Morgan and the late David Perry, whose scholarly genealogical research has not only enabled me to identify all the letter writers and place them in context, but has also provided nuggets of family history which have fleshed out the bare facts of the family tree. Further detail has been provided by Paul Frame, initially through his excellent biography of William's uncle, Richard Price, and then through his generosity in sharing his research material. His encouragement has been wonderfully sustaining, and his creation of the Richard Price Society has introduced me to fellow enthusiasts whose knowledge and support have been invaluable. I send Paul and all the members my very warm thanks.

The archive of the Equitable Assurance Society, now housed at the Institute of Actuaries at Staple Inn, is a rich seam of material relating to William's fifty-six years as an actuary. David Raymont, the librarian at the Institute, has spent many hours patiently guiding me through their records, minutes of meetings, and other documents, ever tactful about my slow grasp of actuarial procedures. As well as a catalogue of the archive material, he provided me with details of eighteenth-century policies and policyholders, giving me a valuable insight into the mores of the time. He also arranged expert help on actuarial science by giving me an introduction to David Forfar, with whom I have had an extended tutorial by means of an exchange of emails. I am indebted to David Forfar for his lucid explanations of the science of big numbers and other concepts, and for his historical accuracy concerning the development of actuarial science.

My very limited science education meant that William's electrical experiments presented me, not to mention those who undertook to coach me, with an enormous challenge. I am immensely grateful to Peter Midgley who gave me a crash course in basic electricity, and to Tony Carrington who gave me step-by-step teaching to explain X-rays and, in particular, William's experiment. John Tucker has continued the tuition and much, much more, being generous with his time and endlessly patient. I could not have completed this book without his help.

William's varied interests took me to a wide range of places for research. I had help and encouragement everywhere I went and I send my thanks to the very many people who gave me their time and the benefit of their expert knowledge. Janet Payne checked all the apprenticeship records at the Worshipful Society of Apothecaries as well as giving me a tour of the impressive Apothecaries' Hall. The librarians at the Wellcome Trust helped me to fill in the details of William's life as a medical student. At the Royal Society I was shown William's original X-ray paper alongside the contemporary letters and papers which gave a fuller picture of the scientific ideas of the time.

I had many happy and useful trips to the British Library, where help was always on hand, similarly at the Glamorgan Archive, where I was particularly grateful that permission was obtained for me to view William's letters in the Merthyr Mawr collection. The University of Bristol Library provided much background material, in particular records of the state trials. The librarians at Stamford Hill Library went to a great deal of trouble to help me discover what had happened to William's house and, at the Hackney Archive, I was shown maps and a wealth of material relating to the time when the house was a YMCA hostel. Vicky Clubb at the Cadbury Research Library, where the YMCA archive is housed, gave me a number of further leads and a route to some delightful photographs. Rosemary Harden at the Bath Costume Museum helped to date my own photograph of Stamford Hill by examining the clothes and hairstyles of the people in the grainy picture.

Alex Allardyce gave me a comprehensive tour of Newington Green which included a visit to the chapel and, most excitingly, one to number 54 where, as an architect, he showed me how to read detail such as the dado in a first-floor room. At Hornsey, Janet Owen and her archive team at the Hornsey Historical Society made me very welcome; they provided details of the Morgan family vault and showed me where to find it. Then staff at the London Metropolitan Archive found the faculty which confirmed the extraordinary inscription about the removal of Sarah Travers's remains.

Ann Thwaite, biographer of both Philip and Edmund Gosse, gave me helpful information about the links with the Gosse family as well as showing great interest in my letters and pictures. I send thanks to her, also to Jennifer Gosse who showed me further portraits.

The Thomas Lawrence portrait of William was just that – a portrait – until Robin Simon examined it from an art historian point of view, and I am indebted to him for his expertise. His suggestion that it holds a coded message added intriguing detail to William's life and a further link to his friendship with John Horne Tooke. This friendship and the significance of the Horne Tooke memorabilia baffled me until Peter Davis directed me to Freemasons' Hall and subsequently provided much material relating to the Jerusalem Sols. At Freemasons' Hall the curator, Mark Dennis, not only suggested the likely provenance of the regalia but also explained the background to eighteenth-century associations such as Jerusalem Sols.

I send my thanks to Paul Frame and Robert Wynne Jones for very generously allowing me to use photographs and prints from their personal collections. Collecting the rest of the images has taken me online to the Bakken Museum, the Library of Congress and the White House and, in the United Kingdom, to the British Museum, the Cadbury Research Library, the Institute and Faculty of Actuaries, the London Museum, the National Portrait Gallery, the Royal Society and the Wellcome Collection. I thank them all for their help and advice.

As well as help with research I had invaluable support with the business of getting the words on the page. Sarah Duncan and Adrian Tinniswood were both superb teachers in the early stages of writing. Members of the Bristol Women Writers were tactful and constructive critics as William's story progressed, giving encouragement when the narrative floundered. I am hugely grateful to the biographer Midge

Gillies, whose mentoring made all the difference to me and the biography. She helped me with structure and style as well as guiding me to make the work more accessible to the general reader. More recently the team at the University of Wales Press have coped with and answered a steady stream of questions, for which I send them thanks. Finally, John Tucker spent many hours going through the manuscript with me, rigorously checking not only the science, but all the research. I am ever grateful to him for his help and for his friendship.

The writing process was not without technical hitches, frustrations and occasional panics. The staff at Far Point and most particularly Paul Hale were quite remarkable in the face of each disaster, calmly restoring lost words to the screen. Pictures presented fresh challenges, but Paul Jones of Mail Boxes Etc. performed magic with the often fuzzy and faded material with which I presented him.

Writing the acknowledgements is in many ways the best bit, not just because it signals the end of a long journey but also because it is an opportunity to remember all the people whose generosity with their time and expertise have contributed to William's story. But it is hard to give adequate thanks for all the advice, the patience, the kindness I have encountered – nowhere more so than with my family, all of whom have showed continued interest through the long years of research. My brother, Simon Bruton, gave me unending support, and his untimely death robbed me of an enthusiastic reader and, more especially, a very dear brother. And, throughout the years, my husband has weathered my highs and lows with wonderful forbearance. He has driven me many miles in pursuit of research material. He has discussed each problem I've encountered. Amazingly, he has kept smiling, and it is thanks to his support that I have kept going and reached the final full stop.

LIST OF ILLUSTRATIONS

Front cover: Feature of early nineteenth-century life assurance policy of the Society for Equitable Assurances, shown with permission of the Institute and Faculty of Actuaries (RC 000243); all rights reserved.

William Morgan

THE WILLIAM MORGAN FAMILY TREE

1. Mary
GIBBON

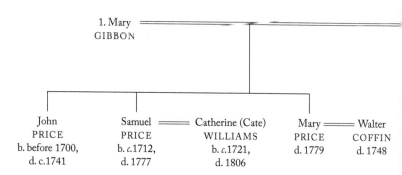

| John PRICE b. before 1700, d. c.1741 | Samuel PRICE b. c.1712, d. 1777 | Catherine (Cate) WILLIAMS b. c.1721, d. 1806 | Mary PRICE d. 1779 | Walter COFFIN d. 1748 |

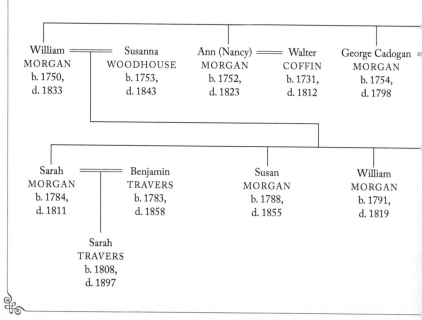

| William MORGAN b. 1750, d. 1833 | Susanna WOODHOUSE b. 1753, d. 1843 | Ann (Nancy) MORGAN b. 1752, d. 1823 | Walter COFFIN b. 1731, d. 1812 | George Cadogan MORGAN b. 1754, d. 1798 |

| Sarah MORGAN b. 1784, d. 1811 | Benjamin TRAVERS b. 1783, d. 1858 | Susan MORGAN b. 1788, d. 1855 | William MORGAN b. 1791, d. 1819 |

Sarah
TRAVERS
b. 1808,
d. 1897

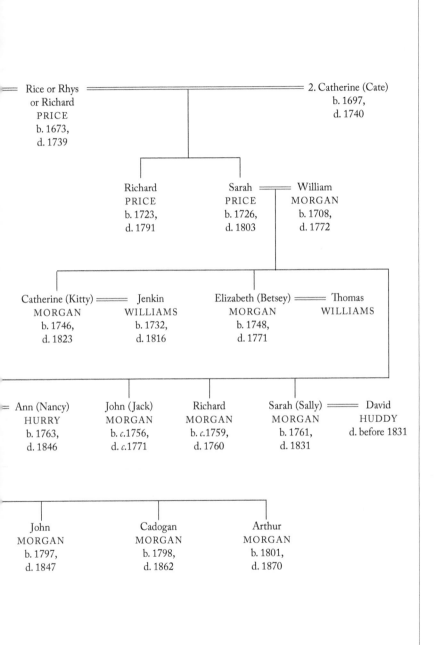

Rice or Rhys
or Richard
PRICE
b. 1673,
d. 1739

2. Catherine (Cate)
b. 1697,
d. 1740

Richard
PRICE
b. 1723,
d. 1791

Sarah
PRICE
b. 1726,
d. 1803

William
MORGAN
b. 1708,
d. 1772

Catherine (Kitty)
MORGAN
b. 1746,
d. 1823

Jenkin
WILLIAMS
b. 1732,
d. 1816

Elizabeth (Betsey)
MORGAN
b. 1748,
d. 1771

Thomas
WILLIAMS

Ann (Nancy)
HURRY
b. 1763,
d. 1846

John (Jack)
MORGAN
b. c.1756,
d. c.1771

Richard
MORGAN
b. c.1759,
d. 1760

Sarah (Sally)
MORGAN
b. 1761,
d. 1831

David
HUDDY
d. before 1831

John
MORGAN
b. 1797,
d. 1847

Cadogan
MORGAN
b. 1798,
d. 1862

Arthur
MORGAN
b. 1801,
d. 1870

PROLOGUE:
THE TEA CADDY LETTERS

As soon as we die we enter into fiction ... Once we can
no longer speak for ourselves we are interpreted.
(Dame Hilary Mantel[1])

William Morgan is my great-great-great-grandfather. A black-and-white engraving of his portrait by Thomas Lawrence used to hang in the gloomy dining room of a maiden aunt.[2] As a child I thought he looked dour and dull. I was wrong.

There is enough in the public domain to paint a picture of a man whose many achievements have affected the way we live today; a man with a sharp wit, a keen mind and strong opinions, and a courageous man who lived in turbulent times. But my introduction to William Morgan colours the canvas in unexpected ways. It came through a legacy: family portraits and drawings, an autograph album and a mahogany tea caddy containing a clutch of family letters. The ancestor I met through his personal letters is just as impassioned and impatient as the public man, but also kind and thoughtful – a loving husband and a devoted father.

In the letters he signs himself 'Will Morgan'. Other letters show that, as a boy, he was Billy.[3] I have chosen to call him William rather than Morgan throughout this biography. Since there are a lot of Morgans (and more than one William Morgan) in the narrative, it makes for greater clarity as well as for a less formal tone. It is also appropriate to my encounter with (and revelations about) his private,

FIGURE 1 William Morgan Esq., FRS.
Engraving by C. Turner (1830), after
a painting by Sir Thomas Lawrence.

domestic world which is as much a part of William's story as his public
life. Through the tea caddy letters I met William's family, happy and
united, but between the lines there are signposts to a sad story that
runs parallel with William's successful life. Even the autograph album,
an innocent collection of signatures pasted on to pastel pages, yielded

some surprising revelations. All biography is to some extent detective work; in William's case it is part of the story, not least in uncovering the risks he undertook.

When William was born in 1750 the heads of the 1745 Jacobite rebels were still impaled on poles at Temple Bar – a reminder that religious affiliation was still part of politics. Roman Catholics and Dissenters were barred from public office. The king chose his ministers from the Houses of (hereditary) Lords and Commons. Seats in the Commons could be bought with bribes and threats. The system was flawed and open to corruption.

William was a Dissenter and a reformist. He mixed with the radical thinkers of the day, amongst them Benjamin Franklin, Joseph Priestley, Richard Price, John Howard, Thomas Paine, John Horne Tooke and Francis Burdett. Through his membership of reformist societies and through his own publications, William campaigned for electoral reform and government accountability. In doing so he took colossal risks and narrowly missed being sent to the Tower of London.

Like many of his contemporaries, William was fascinated by the science of the day and in particular by electricity. He conducted experiments to determine how electricity 'worked' – experiments with hazardous materials which were not without risk. When his younger brother, George, died aged only forty-four, the family blamed his death on poisonous fumes inhaled during electrical experiments. William, who nursed his brother through his final days, was well aware of the dangers of his experiments.

As for William's fifty-six years at the Equitable Life Assurance Society (SEALS),[4] here he was dealing with matters of life and death but at one remove. At the Equitable William learnt how to understand and manage financial risk. In 1789, for his work on the mathematics of life assurance, he was awarded the Copley Medal, the Royal Society's most prestigious decoration. Subsequent generations have hailed him as the 'father of the actuarial profession' – recognition of his having established many of the rules and standards on which the science is based.

The picture on the front cover of this book is the motif used from 1800 to 1899 by the Equitable on its policies. It is something which

William would have seen every day and I like to think that he might have approved the design. In the course of his tenure as Actuary, the Equitable became one of the most successful insurance societies of its time. Its success continued under William's son, Arthur Morgan, and by the twentieth century the Equitable was the dependable insurance company of choice for many professional people. Its problems in the 1990s and its demise in 2000 was a shock to the financial world.[5] This is William's story, not the Equitable's, so I shall not be examining the reasons for its failure except to say that, had the Society stuck to William's rules of prudent management, the crash could have been avoided. He would certainly have been devastated by its ignominious end.

William gets a mention, and due praise, in works on the history of actuarial thought. He has a cameo role in the autobiographies of his great grandson, Arthur Waugh,[6] and his great-great-grandson Evelyn Waugh.[7] He gets a significant part in the numerous biographies of his uncle, Richard Price, the most recent of these being *Liberty's Apostle*, by Paul Frame.[8] He also appears in *Travels in Revolutionary France*, the edited letters of his brother, George, from Paris in 1789.[9]

The fullest account to date of William's life is given in *A Welsh Family From the Beginning of the Eighteenth Century*. The author, Caroline Williams, was William's great-niece. There is no record of their having met (and she was only nine when William died) but she would have had first-hand accounts of the Morgans from several members of the family, in particular William's eldest granddaughter, Sarah Travers. Sarah Travers grew up, from the age of three, in William's household as his ward, effectively a daughter. As well as her knowledge of the family Sarah Travers had access, after William's death, to Morgan papers, some of which she bequeathed to Caroline Williams.[10]

Caroline Williams, writing in 1893, gives a panoramic picture of the Morgan family over two centuries. It is a remarkable work about a remarkable family and introduces a huge cast of characters. Whilst some of her material is obviously based on anecdotes passed down through the family,[11] most is taken from sources which she lists in her Preface, in particular 'the letters of three generations' and 'notes for a biography' of William's uncle, Richard Price, written by the poet, Samuel Rogers. In addition she quotes from William's diary. Sadly much of this source material is lost (probably destroyed). My tea caddy legacy, just a fragment of the original, contains letters which Caroline Williams used and from which she gives accurate quotations – but she makes mistakes. The digitised edition of her work shows some handwritten amendments made by the owner of the copy used. My own original 1893 edition once belonged to one of William Morgan's granddaughters[12] and has her pencilled corrections and comments in the margin, which, for example, point out spelling slips and telescoped dates. More seriously (but with no subsequent margin comments) Caroline Williams sanitises the story of William's guardianship of Sarah Travers; by being selective in the material she uses, she omits any mention of the feud with his son-in-law. I have learnt to treat her account with caution.

Even the date of William's birth poses questions. A handwritten family tree in my 1893 edition of *A Welsh Family* gives it as 26 May 1750. The same date is given by the distinguished actuary Sir William Elderton in an address to the Faculty of Actuaries in 1931[13] and in the current Oxford Dictionary of National Biography. But the date inscribed on William's tomb is 6 June 1750. The probable reason for this discrepancy is discussed in chapter 1.

Caroline Williams (and later Sir William Elderton) note William's friendship with the radical thinkers of the day, in particular John Horne Tooke, but nothing in her account explains a puzzling memento in the tea caddy legacy – a slender cardboard box labelled 'Horne Tooke's ribbon and lind stone'. Inside, amidst tissue paper smelling strongly of mothballs was a wide red sash, on which was sewn something best described as a rosette (see Figure 21). A lind, according to the *OED*, is a shield, but this, unless it has talismanic properties, is decorative rather than protective: a round ceramic disc about four centimetres in diameter surrounded by a circle of brilliants and set off by a frill of pleated blue silk. On the central disc is a picture of a woman, perhaps a goddess, crowned and carrying a cornucopia of flowers. The whole effect is bright and flashy, what today might be scorned as bling. When and where would Horne Tooke, politician and radical thinker, have wanted to wear such a garish ornament, and why did William treasure it? The question became even more puzzling when I discovered, beneath the tissue-paper nest, a scrap of folded paper – a subpoena summoning William to the trial for high treason of John Horne Tooke. The 1794 treason trials were a legal landmark, and the subpoena indicates that William was more than one of the many interested bystanders. Its preservation in the same box as the lind stone begs the question of how the two might be linked. Solving these mysteries was to take some sleuthing.

No detective work, however, was needed to discern William's love of Wales, which shines through his letters. He returned to Bridgend whenever he could, and most summers took a house – Greengate – in the village of Southerndown. In a letter of 1796 to his younger daughter, Susan, he tells her about a trip to the beach where the party, nineteen strong, 'dined upon the rocks' – what a grand way to describe a picnic!

In the same letter he describes children's races on the sands. He is very modern in his attitudes to parenting – and in other ways. William lived in the eighteenth and early nineteenth centuries but his opinions and wisdom belong as much to the twenty-first century. He would undoubtedly have been interested in climate change, genetically modified crops and livestock, artificial intelligence, and more. He had strong views and was never afraid of expressing them, never afraid of criticising the government even when, in doing so, he risked imprisonment. During the Napoleonic wars he despaired of the government's financial incompetence and the damage being done to the country's economy. Knowing him as I do after more than fifteen years in his company, I feel sure he would have approved of the EU – criticised it certainly – but argued passionately and cogently against Brexit. And what a joy it would be if we could hear on television and radio his uncompromising opinions expressed in colourful language and spoken with a lilting Welsh accent.

Notes

1. Dame Hilary Mantel, *The Day is for the Living* (Reith Lecture, 2017).
2. The original oil painting now hangs in the Great Hall of the Institute of Actuaries at Staple Inn.

3. Richard Price refers to his nephew as Billy in a letter of 17 June 1770. See Caroline E. Williams, *A Welsh Family From the Beginning of the Eighteenth Century* (London: Women's Printing Society Ltd, 1893, repr. Kessinger Publishing Legacy Reprints), pp. 34, 36 and 51.

4. The full name is Society for Equitable Assurances on Lives and Survivorships (often abbreviated to SEALS), see p. 59.

5. For an outline of the demise of the Equitable see 'Equitable Life: Timeline of key events', on the BBC News website. Available at *https://www.bbc.co.uk/news/business-10725923*. Accessed 2017.

6. Arthur Waugh, *One Man's World* (London: Chapman and Hall, 1931), pp. 11–12.

7. Evelyn Waugh, *A Little Learning* (London: Chapman and Hall, 1964), pp. 10–11.

8. Paul Frame, *Liberty's Apostle* (Cardiff: University of Wales Press, 2015); see p. 305 for index of references.

9. George Cadogan Morgan and Richard Price Morgan, *Travels in Revolutionary France & A Journey Across America*, ed. Mary-Ann Constantine and Paul Frame (Cardiff: University of Wales Press, 2012); see p. 236 for index of references.

10. The specific bequest in Sarah Travers's will is: 'I give to my cousin Caroline Elizabeth Williams of Vicarage Gate Kensington the ring containing the hair of Dr Franklin and Dr Price and Dr Prices letters manuscripts and correspondence also the twelve china plates scroll pattern formerly belonging to him.'

11. See for example Dr Morgan's remarks about Jenkin Williams (p. 4) and the exchange between William and Richard Price (p. 25).

12. The granddaughter is Mary Susan Morgan (1824–1921), eldest daughter of William's second son, John Morgan (1797–1847). Mary Susan Morgan is my great-grandmother.

13. William Palin Elderton, 'William Morgan, F.R.S., 1750–1833', in *Transactions of the Faculty of Actuaries*, 14 (1931–4), 1.

1

BRIDGEND

May I that pleasing seat attain
Where Ogmore's waters roll,
Where rapid Ogmore swiftly glides
And flocks adorn its flowery sides
To raise my drooping soul.
(William Morgan, *Ode in imitation of Horace*[1])

It is 26 May 1750. At the top of Newcastle Hill in Bridgend is the doctor's house. Inside, the doctor's wife, attended by a trusted midwife, is in labour. This is her third confinement. She already has two daughters so it isn't hard to imagine the mother's joy when the midwife tells her that this time the baby, crying lustily, is a boy. Her husband is hoping for a boy – a son to carry on the family name and in due course to take over his father's medical practice. The baby, his mother knows as she takes him in her arms, already has his life mapped out for him. He is tightly wrapped so she does not yet know about his deformity. The midwife, and later the doctor, shield the truth from her in the first hours after her labour. They allow her to rest, to suckle the child and marvel at the tiny features of his face.

It is tempting to picture the scene in this way. It can't have been long before the swaddling shawl fell away and Sarah saw baby William's foot. A strange misshapen thing, more like a ladle than a foot. Perhaps she thought he had bent it in the process of being born. Perhaps she assumed that the distorted limb would straighten with time. It would, presumably, have been her husband who gave her the name of their son's condition: congenital talipes equinovarus. He had a club foot.

It is even possible that the birth gave cause for concern rather than celebration; possible that William was a sickly child, his survival so much in question that he was given a hasty baptism by the midwife and the subsequent registration overlooked. That might account for the fact that, whilst each of his seven siblings was baptised at Coity church, no record of a baptism of William has been found.

An entry on the baptismal register might have helped to confirm William's date of birth. Handwritten records by family members give it as 26 May 1750, but the date inscribed on his tomb is 6 June. How did this discrepancy arise? The clue would seem to be in the number of intervening days – eleven. When the Gregorian calendar replaced the Julian calendar in 1752, eleven days were removed from that year so that 2 September was followed by 14 September. William at two years old would have been too young to join in the popular outrage at the apparent theft of days from his life, but he might, in later years, have needed to make adjustments in some of his actuarial calculations.

Whatever his date of birth, and whatever the odds against a healthy life might have been, club-footed William survived. Today a club foot can be corrected either with surgery, the preferred procedure for much of the early twentieth century, or with a series of plaster casts, each moving the foot very slightly until the bones lie in a normal position. This non-surgical method is a return to a much earlier approach when the foot was tightly bandaged, sometimes splinted, into position. We do not know how William's foot was treated but subsequent events show that his disability was visible. It seems likely that he had to wear a special boot and that he was destined always to walk with a limp – a disability which was to have a profound effect on his life and his career.

William's parents could not protect him from the public recoil, the taunts and teasing about his clumsy gait which he was likely to encounter at school and in the world of work, but they could – and did – give him a secure and loving family circle. For his sisters, four-year-old Kitty and two-year-old Betsey (no-one called them Catherine and Elizabeth), he was first 'the baby' and then, when further babies arrived, he became Billy. By 1759 there were five more Morgan children: Ann, always known as Nancy, George, Sarah, known as Sally, Richard (who died at the age of one) and Jack.

The Morgan girls were educated at home; the boys probably attended Cowbridge Grammar School, reputed to be one of the best schools in Wales, where the students received a classical education. George was certainly a pupil, perhaps a weekly boarder, and in due course became head boy. Surprisingly, William does not appear on the pupil records, though it is likely that he also went to the school. For a start it seems unlikely that the eldest of the brothers would not have been educated there, and in addition there is evidence that William was well versed in the Classics. When he was only nineteen, he composed some verses in imitation of *Ode of Horace*, Book 4.[2] The poem celebrates the 'peace of mind' that a return to his birthplace can give him and is addressed to 'Dear George', showing the fondness of the two brothers for each other. In fact, William had good reason to envy his younger brother: not only was George clever, he was strikingly good-looking; William, with his deformed leg, was awkward and, as shown in the two adult portraits of him, had a rather plain face. George was allowed the freedom to choose his studies. He matriculated in October 1771 and went up to Jesus College Oxford to study Classics.[3]

It was different for William. His career was already decided; he was to become a doctor. No one seems to have anticipated the problems his club foot would bring. Medicine was what his father wanted for him, although he admitted in a letter to his wife's brother, Richard Price, that his eldest son had 'a greater inclination for academical learning than for the study of pharmacy'. Dr Morgan did not bend: 'Billy is willing to comply with my wishes, notwithstanding'.[4]

Life was less serious for William's brothers and sisters. Although the girls were expected to study, they preferred parties and dances; Kitty and Betsey were consumed with envy when Nancy had a season in Cowbridge – much more sophisticated than Bridgend. So, not surprisingly, there were gaps in their educational achievements. Kitty's spelling was so bad that she had to beg George to help her with her love letters. When George felt he had better things to do and refused, Kitty locked him up until the letter was completed. George took the

FIGURE 2 William Morgan. Engraving by William Say (1803),
after a painting by George Hounsom.
(By kind permission of Paul Frame)

obvious means of revenge by deliberately including misspellings. His
ruse, however, did not do too much damage, for Kitty eventually married her beau, despite some initial reservations from the family. 'Girls,'
their father warned, 'don't lose your hearts to this young coxcomb with
his fine gold-laced hat.'[5]

The anecdotes paint a cheerful and carefree picture of family life.
Dr Morgan was held in affectionate respect throughout Glamorganshire
as he rode from one end of the county to the other with his medical

book in his saddlebag. In the kitchens of the houses he visited he prepared the medicines of the day; at the bedsides of the sick he dispensed common sense and kindness.

In 1744, six years before William's birth, Dr Morgan attended two sisters who had recently lost both parents and who were suffering from a mild but persistent fever. He prescribed a better diet as the best cure for their illness. The girls, aged eighteen and sixteen, were the granddaughters of a Dr Richards, another popular doctor and the one from whom Dr Morgan had taken over the practice. Perhaps it was this connection that made him especially attentive to the sisters, for he quietly arranged for them to have food from his own kitchen. More likely he was already in love with the older girl, Sarah Price, and before long she came to love her handsome doctor. He was thirty-five, a childless widower and almost twice her age, but in other respects they were well matched: clever, good-looking and each from a long line of Welsh gentry.[6]

Dr Morgan claimed an impressive Welsh pedigree, tracing his lineage back to one Cadogan Fawr who, in 1294, had led a war band against an invading English force. Legend has it that, having killed many of the foe and seen off the remainder, Cadogan refused to sit back and celebrate. 'Hoeg fy mywall', he called out to his henchman, 'Sharpen my axe'. It became the family motto.[7]

Dr Morgan had a small family estate, Tyle-coch.[8] Sarah's family owned much more land but it had been unevenly divided when her father died, leaving his wife and her three children almost destitute whilst a son by a previous marriage enjoyed great wealth. Sarah Price was poor, a dowry-less bride, when she and Dr William Morgan were married in Llandaff Cathedral on 7 December 1744. After the ceremony; she mounted his horse and rode pillion behind him back to Bridgend.

Bridgend today has a population of nearly 50,000, but in the eighteenth century there were no more than 1,000 inhabitants.[9] Dr Morgan was well educated, witty and charming; his circle, both professional and social, went way beyond the town. It was as a dinner guest as well as a doctor that he visited Dunraven Castle and many of the smartest homes of the county.

But for all his popularity and success, Dr Morgan was no business-man and he was not rich. He was generous to the poor and honour-able in his treatment of the wealthy. When one of his grateful patients promised to make him his heir, Dr Morgan stood to inherit the pretty hamlet of Merthyr Mawr. He refused, reminding the squire, 'you have a nephew of your own'.[10]

Had Dr Morgan charged higher fees for his services, there might have been enough money to pay for a decent medical training for his son; instead, William suffered considerable financial hardship in his medical apprenticeship. He was just nineteen when he said goodbye to his family and set off for London. There is no record of his journey but, whatever his route and mode of transport, he had at least three, probably four, days' travel, slow and uncomfortable, ahead of him.[11]

Many of the roads were only dirt tracks, their soft surfaces deeply rutted by the wheels of the wagons and coaches, making for a rough ride. In wet weather they quickly became a quagmire and were some-times impossible to use. William, setting out in the summer of 1769, probably began his journey in a carrier's wagon – a long vehicle covered with a hood and drawn by a team of eight horses. These cost only a half-penny a mile but they were slow, covering little more than two miles an hour. He might have travelled via Gloucester, where there was a bridge across the river Severn, or he might have made his way to Bristol, cross-ing the river by ferry. Thereafter, he could take a stagecoach but, with thin metal wheels and no springs, they were hardly more comfortable than the wagons. And they were expensive – 2d (twopence) or 3d a mile to sit outside, 4d or 5d for an inside seat and, on top of the fare, the coach driver would have expected a tip. With both coach and wagon there were the additional expenses of food, drink and overnight accom-modation, so that the trip to London was a major outlay and William could not hope to return home for many months.

As travellers neared London and they passed through the lit-tle villages of Kensington and Knightsbridge, the traffic increased. They shared the road with cattle, sheep and pigs being driven to the abattoirs in the city and to the market at Smithfield, their dung add-ing to the mud, household rubbish, dead cats and other detritus in the path of the coach. The market gardeners in the area paid a good price for the London street dirt and for the contents of the domestic cesspits collected after dark by the night-soil men, but the trade was not enough to keep the streets clean or to prevent the all-pervading rank smell.

It was not just the smell – the noise, the bustle, and the sheer size of London were a shock to anyone arriving there for the first time. The population had already spilled out beyond the confines of the city. John Rocque's map published in 1747 shows the developments with names still familiar today: Cavendish Square, Grosvenor Square and Berkeley Square, all approached by a neat arrangement of surrounding streets.[12] By the end of the century many more houses had been built. It is difficult to give an accurate figure for the pre-census population of London but it is generally estimated to have been about 650,000. William's home town of Bridgend was small enough for him, as the local doctor's son, to know personally most who lived there.

Some of the coaches from the West Country stopped at one of the several inns in Piccadilly; some went as far as the Swan with Two Necks in Lad Lane, Wood Street. Wood Street, just west of the Guildhall, still exists, but Lad Lane and, with it, the Swan with Two Necks have long since disappeared beneath office blocks. A coach going as far as Lad Lane would have taken William along the Strand and into the City through the arch at Temple Bar, and beneath the two remaining Jacobite heads impaled there since 1746 and by now black shapeless lumps.[13] Next, along Fleet Street, past St Paul's Cathedral, completed in 1711, dominating the skyline but already becoming grimy with smoke pollution.[14] As he travelled deeper into London, William would have seen sedan chairs, their occupants being carried clear of the rabble as well as the filth underfoot.

Anyone arriving in such an unwelcoming place was bound to feel lonely, homesick and fearful. William had one place where he could seek comfort and he made straight for the home of his mother's brother, Richard Price, who was sympathetic and welcoming. He was especially fond of all his nephews and nieces; they were almost a surrogate family – his wife, Sarah, was an invalid after a serious illness in the early days

FIGURE 3 Richard Price, DD, FRS. Engraving by
Thomas Holloway (1793), after a painting by Benjamin West.
(National Portrait Gallery)

of their marriage, and unable to have children. By the time William
came to stay with them, Price was forty-six and Sarah forty-one. They
were quite comfortably off and living at Newington Green, then a small
village to the north of the capital. It is part of London now, just a thirty-
minute bus ride from King's Cross station. The automated voice tells you

FIGURE 4 Home of Richard and Sarah Price in Newington Green,
today number 54. It is the property with the squared-off top.
(By kind permission of Robert Wynn Jones)

when you have arrived, and there it is: a little square of green – grass,
shrubs, trees – surrounded by houses.

Eighteenth-century travellers would have thought little of trudging
the two or three miles from the Swan with Two Necks to Newington
Green. Given his club foot and even a modest bag of luggage, William
might have paid the 4d fare for one of the twice-daily coaches from
the City taking him through green fields to an inn at Spring Gardens,
from where he had only a short walk to his uncle's house. Today it
is number 54 and stands, the central house on the west side of the
Green, in what is now the oldest terrace in London. Built in 1658, it is
a handsome red brick house with large sash windows; inside, the original staircase winds up the five storeys from cellar to attic leading to
spacious rooms, each with a fireplace. In one, on the first floor, the dado
remains and suggests that the room was used as the dining room, for the
wooden panelling is at just the right height to have protected the wall

FIGURE 5 Newington Green Chapel as it is today.
(By kind permission of Robert Wynn Jones)

when chairs were pushed back from the table. The large garden to the rear of the house has long since been nibbled away by infilling, but the view from the front windows is similar to that of 1769: a small green enclosed by buildings.[15] The original houses have been replaced, most obviously on the south side, which is now dominated by Hathersage Court, an unexciting block of flats built in 1970, but there, on the north side, is the chapel where Richard Price was minister and which still has a thriving congregation.

His uncle and aunt made it clear that William could stay with them for as long as he liked. It must have been tempting.

NEWINGTON GREEN

Hail, MEMORY, hail! In thy exhaustless mine
From age to age unnumbered treasures shine!
(Samuel Rogers, *The Pleasures of Memory*)

Next door to Price lived Thomas Rogers and his family, and the memories of Samuel, one of his sons, provide a picture of life at Newington Green. Samuel was only six when William arrived but by the end of the century he was a well-known poet. His poetry is no longer fashionable but in the eighteenth century it was widely read and much admired. Success and inherited wealth meant that he moved in exalted circles and entertained people of rank and influence. His diaries and his table-talk records still provide insights into the politicians, writers and actors of his day. In their turn they provide a view of the adult Samuel Rogers: they were happy to accept his hospitality but he was regarded with suspicion, even dislike, by those who knew him. Sir Walter Scott said of him: 'It matters not what ill we say of Rogers behind his back, since we may be pretty certain that he has said as much of us behind our backs.'[1]

Even his detractors, however, acknowledged that he had a very good memory and vouched for the accuracy of his diaries and of the careful records of the sharp and witty repartee at his dinner table. And his actions belied his words. The actress, Fanny Kemble said of him that he had the 'kindest heart' although the 'unkindest tongue' of anyone she knew.[2] His poetry too, although clever with technical polish and mannered showiness, is gentle in tone and has none of the bite of his

FIGURE 6 Samuel Rogers.
Chalk drawing by George Richmond (1848).
(National Portrait Gallery)

Table-Talk reminiscences. He seems a reliable witness and his memories of his Newington Green childhood are fond and touching.

He remembered with particular affection evenings when Richard Price left his study – informal in just his dressing gown – and joined the Rogers family who lived next door. 'He would talk and read the Bible to us till he sent us to bed in a frame of mind as heavenly as his own.'[3] On Sundays Samuel attended chapel and listened to Price's sermons, which impressed the young boy so much that for a time he wanted to become a preacher himself.

By 1769 Thomas Rogers, a banker, and his wife, Mary, had six children, ranging in age from eight to a one-year-old toddler. (Five more children were born in subsequent years.) The children were younger than William's brothers and sisters but, with them, he could enjoy the family life he was missing, and Samuel Rogers's later reminiscences record the stories with which the children entertained William. Price had been discovered, they told William, one summer's day retracing his steps on a walk across the fields to rescue a beetle which he had noticed lying on its back and struggling to turn over. On another occasion Price had been seen leaving little piles of coins in a field near his house. When asked what he was doing, he had explained that he had freed some larks from the nets set to trap them and then felt guilty when he realised that each liberated lark was a loss of income for some unknown person. He trusted that the 'right' person would find the money he had left beside the empty and damaged nets.

These charming stories reveal a warm, if eccentric, side to Price – the revered polymath, whose wide interests included mathematics, science, politics, theology and literature. When William arrived at Newington Green, Richard Price and Thomas Rogers had a weekly supping club where discussion and debate were as important as the meal. As a guest, William would have been exposed to a heady mix of radical political and intellectual thinking.

And theology. Price, Rogers and other members of the supping club were Dissenters. Dissenters can be broadly defined as those unable to accept the Thirty-nine Articles of the Church of England. The Tudor and Cromwellian purges were over, but religious differences had left a legacy of divisions in society. The 1745 Jacobite rebellion, only twenty-five years previously, was a reminder of the see-saw power struggle between Roman Catholics and Protestants. Roman Catholics were kept in check by the Test and Corporation Act (1661), which denied public and military office to those who refused to take religious sacraments at an approved parish church. They were excluded by the Act from Oxford and Cambridge universities, as were Dissenters. During the years of the Commonwealth under Oliver Cromwell, Anglicans were deprived of political power, denied prayer-book worship and burdened with extortionate taxes. When the monarchy was restored and the Church of England re-established, Anglicans took revenge and imposed strict limitations on the lives and careers of Protestant Dissenters. They were outsiders and, like Roman Catholics, denied a number of rights and privileges. The term 'Dissenter' was something of a catch-all label, for doctrinal differences varied from the austere and puritanical to the liberal and compassionate. Price was a Dissenter but his was, in his own words, 'not a sour or enthusiastical religion'.[4]

William's respect for his uncle's approach to religious debate is clear in his *Memoirs of the Life of Richard Price* (1815) in which he records the differences in metaphysical thinking between Richard Price and Joseph Priestley.[5] At Price's suggestion Priestley published the correspondence relating their debate. William leaves the readers of the *Memoirs* to examine the published discussion for themselves but notes that the disagreement 'neither disturbed the friendship of the parties, nor abated the high opinion which each entertained of his adversary's talents and integrity'. It is an interesting insight into the two men's friendship, but he does not leave it there. With a touch of understated humour he adds 'like all other controversies of the same kind, it left both as it found them, in the full persuasion of the truth of his own opinions'.[6]

Elsewhere in the *Memoirs* he records another disagreement, that between Price and David Hume. The two men differed fundamentally in their philosophical views, but each had respect for each other's academic rigour. According to William, on one of David Hume's visits, Price 'succeeded in convincing him that his arguments were inconclusive',[7] but letters between the two men, though courteous and candid, give no clues about the precise point of their discussion.[8] Why was William so coy about giving details? Perhaps an excess of courtesy, perhaps part of a reticence to discuss his own beliefs, or perhaps he wanted to tease his readers.

The *Memoirs* give an outline of Richard Price's early life and reveal many similarities between Price's situation and his nephew's. Price had, in 1740 and aged only seventeen, come to London from Bridgend. He was poor; his half-brother lent him a horse but only to get him as far as Cardiff; after that he had no choice but to walk and hitch lifts whenever he could. Once in London he was able to continue his education, thanks to the generosity of an uncle. Price was willing to give his nephew similar financial support whilst he undertook his medical apprenticeship, but William did not want to take advantage of his uncle's kindness. Instead, he set about apprenticing himself to an apothecary.

Was it haste, or ignorance, or even misplaced pride which led William to embark on his apprenticeship to Mr Smith of Limehouse Docks? It was a very dodgy appointment and one that William quickly regretted. For a start, Smith was only a self-styled apothecary. His name is not included in the records of the period at the Worshipful Company of Apothecaries, membership of which, by then, was on a professional basis.

Apothecaries had come a long way from their early beginnings as grocers administering drugs alongside spices, perfumes and sweetmeats. They still ranked low in the medical hierarchy; physicians had long been – and remained – at its head, but they were scarce and expensive. For the poor and those of modest means it was simpler and cheaper to turn to the apothecary. Though granted a charter in 1617, the apothecaries had for over two centuries no legal right to claim a fee. They could, however, charge for dispensing remedies, so advice was usually accompanied by a potion or a powder. Gradually they became known as doctors and, with a House of Lords ruling in 1704 giving them the right to practise, they became the forebears of our general practitioners.[9]

Smith, however affordable his drugs, should not have practised without serving the seven years' apprenticeship which entitled him to become a Yeoman of the Company. It is unsurprising that he practised in Limehouse Docks, which was not a salubrious area and where the patients were poor dock labourers. William was expected to work long hours and, at the end of the day, to sleep under the counter.

'He treated me no better than a dog', William recorded in his diary.[10] He stuck it for three months until his 'Welsh temper' could stand it no longer', and he laid Smith in the gutter. William's Welsh temper was to re-emerge at key moments throughout his life and, with this early moment of anger, he put an end to his Limehouse Docks apprenticeship. Perhaps he had spent the three months planning his moment of

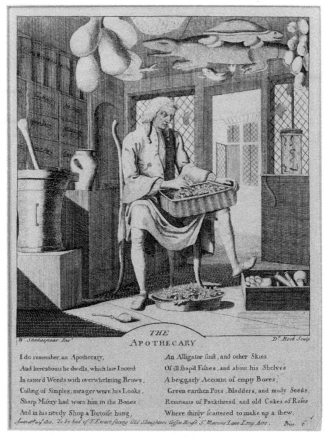

FIGURE 7 An apothecary sitting in his shop, sorting through materia medica, surrounded by paraphernalia of his profession. Engraving by Dr Rock (*c*.1750), after W. Shakespeare.
(*The Wellcome Collection CC BY*)

sweet revenge, because he had timed his flash of rage very conveniently for the last day of his first quarter and he had made advance preparations. On the following day, 11 October 1769, he was apprenticed to a new master, Joseph Bradney, in Cannon Street.

Price helped to arrange the new appointment. By this time William seems to have been ready to accept financial help from his uncle in order to meet the annual fee of £16, and the further expense six months later when, in May 1770, he entered St Thomas's hospital as one of the pupils and dressers. Dressing pupils paid as much as £50 for the privilege of changing bandages and attending to wounds. They were also allowed to perform minor operations, such as the commonly prescribed bleeding, and to give general assistance to the surgeon in his work.

In addition there were lectures, at which William took careful notes in his neat copperplate. Thomas Smith, the eminent surgeon under whom he studied, judged William to be his best pupil;[11] whether this is a comment on his industrious note-taking or on his skill with patients is not clear, but his willingness to work hard set the tone for his entire life.

Working in the wards cannot have been a pleasant task. The rich were nursed at home; it was the poor who went to hospital. Hogarth had died in 1764 but these were still the poor who throng his crowded canvases and engravings: rough and ragged, drunken and dirty. In hospital they would at least get a regular meal and they could send out to a dram-shop for gin with which to wash it down. The gin might have helped them to endure the bedbugs that lurked in the wooden bedsteads and would certainly have been helpful for anyone undergoing surgery. There were, of course, no anaesthetics.[12]

William would have attended hospital in his everyday clothes: a tail-coat over a waistcoat and knee-breeches, and, very likely, a wig under his three-cornered hat. No washable uniforms, alcohol hand pumps and other aids to hygiene for the eighteenth-century medics. It is a wonder that anyone weathered a stay in hospital, and expectations were low – at St Bartholomew's, on admission, patients were required to pay a deposit of seventeen shillings and sixpence (a week's wage for a journeyman tradesman at the time) to cover the burial fee – refunded if you survived.[13]

On top of his hospital work, William had his apprenticeship duties in Joseph Bradney's apothecary's establishment. 'It is as much as I can do to attend the shop and hospital', he wrote to his mother.[14] During the winter months he found lodgings in Southwark, but with the longer days of summer time he could manage the journey from Newington Green and he lived with his uncle and aunt, where, once again, he spent time with the Rogers family and heard more stories about his uncle's entertaining behaviour.[15] And it was good for him, as a distraction from his harrowing work at the hospital, to see his serious uncle in carefree mood. On one occasion the Rogers children persuaded Price to attempt a leap over the honeysuckle bush in their garden. Instead of clearing the bush, he landed in its middle and had to go home entwined with honeysuckle. They all watched with delight when Price challenged Mr Hulton, a commissioner of customs at Boston, to a hopping race. William could not fail to smile at the sight of two grown men, one in clerical black clothing and wig, making their one-legged way across the cowslip meadow.

As well as recording these snapshot memories, Samuel Rogers remembered the excitement of being introduced to the microscope, the telescope and the electrical machine (more accurately, a machine for producing static electricity) which Price had in his study.

Telescopes and microscopes still have sufficient magic for modern children to enjoy, but electrical machines are now only to be seen in museums and on YouTube. The original eighteenth-century contraptions consisted of a globe or cylinder of glass which, when turned on an axle, created friction with a piece of cloth, or leather, or even a human hand, and thus produced a charge of static electricity. It is much the same as the familiar effect which can be produced by stroking a cat for long enough to allow its fur to become charged with static and stick upright, sometimes even crackle and emit tiny sparks. If there is no cat to hand, vigorously combing one's own hair does the trick.

Today it takes a power cut to remind us how much our lives are dependent on electricity. Imagine spending a winter's evening eighteenth-century style with only candles and oil lamps for light, and you have some idea of how thrilling it must have been for the Rogers children – and William – to see sparks flying from Price's electrical machine. They were not alone; electricity was exciting stuff. It was spectacle and it was entertainment.

In London, and in fashionable cities such as Bath, people flocked to see electrical power made visible. 'It is all the vogue', wrote one reporter, 'electricity has replaced the quadrille'.[16] Showmen would not only make sparks fly but, with a sword suitably charged, set alight a

ESSAY SUR L'ELECTRICITÉ DES CORPS · FRONTISPICE

FIGURE 8 An Abbé Nollet electrical session in Paris, 1746.

The electrified boy has been suspended on silk cords. Beneath his left hand
chaff flies, whilst beneath his right hand papers are disturbed. Abbé Nollet
is standing on the left holding an electrified glass tube. The seated woman
on the right is about to discharge the boy's nose with her finger. Frontispiece
from Jean Antoine Nollet, *Essai Sur L'Électricité Des Corps* (Paris, 1753).
(Collections of the Bakken Museum, Minneapolis)

glass of brandy, and perform other unlikely tricks. One of the most famous shows involved suspending a small boy from the ceiling by silken threads. (Boys from charity schools were easy to come by.) After he had been charged by means of an electrical machine the child would attract, for example, feathers or iron filings. Audiences watched in rapturous amazement as the feathers or filings flew from a dish to the boy's outstretched hand. By the middle of the century scientists were able to 'store' electricity by means of the Leyden jar. This was a glass jar partially filled with water. With one end of a brass wire dipped into the water and the other connected to an electrical machine the water became charged with static electricity, and this allowed even more spectacular demonstrations of its power. In France, Jean Antoine Nollet entertained King Louis XV by sending a current through a chain of 180 Royal Guards. King Louis was highly amused when the soldiers all jumped simultaneously into the air.[17]

Natural philosophers – they were not yet known as scientists – could not fail to enjoy discovering and demonstrating the power of electricity, but the public displays were not merely entertainment. They needed to convince people that what they were doing was useful and interesting. We expect today's scientists to analyse and dissect, to measure and calculate, to check and test, and – most importantly – to theorise and predict. Eighteenth-century researchers had to be canny; it was not long since experimentation had been seen as akin to dabbling in black magic.

William, sharing childish pleasure with the Rogers children as his uncle Richard's electric machine produced sparks, was intrigued at a more serious level. He wanted to discover for himself more about electricity, and there was another aspect which interested him: the medical implications. The prospect of healing by means of electricity was as serious a hope at the end of the eighteenth century as stem cell medicine is at the beginning of the twenty-first century, and the archives of the Royal Society give some idea of the astonishing powers ascribed to electricity. In just one year (1785) papers were presented on 'Loss of Speech after Resuscitation from Drowning restored by Electricity',[18] 'Two Cures for Blindness by Electricity',[19] 'Dumbness treated by Electricity',[20] and 'Treatment following a Species of Fever called Typhus Nervosus'.[21] All claimed sensational cures. The seaman who was dumb for a month following a near drowning was treated with an infusion of horseradish in Madeira wine and electrical shocks to his tongue. After two weeks 'he recovered the articulation of his voice as perfectly as he ever enjoyed it in his life'.[22] Nineteen-year-old Sarah Baldwin, following an attack of 'typhus nervosus', 'did not return to her usual bloom of health' but suffered from low pulse and fainting fits – until she had electric shock treatment which restored her to 'the most perfect health'.[23]

It was even hoped that in electricity would be found, quite literally, the vital spark of life. In 1818, only thirty years later, came the publication of Mary Shelley's novel *Frankenstein*, in which Victor Frankenstein, the modern Prometheus of the subtitle, assumes a god-like role in his creation of human life by means of the electricity. Here, in a Gothic wrapper, was contemporary science fiction which shocked and thrilled her readers.

William's ideas about electricity were not as sensational as Mary Shelley's, but his interest was awakened and he had the scientist's desire to push back the boundaries of knowledge. He wanted to experiment, and to learn more about the properties of this powerful force. Some ten years later he was to conduct his groundbreaking experiments now recognised as an early, possibly a first, step in the discovery of X-ray.

3

BLACKFRIARS

The bridge at Blackfriars is a noble monument of
taste and public spirit. I wonder how they stumbled
upon a work of such magnificence and utility.

(Matthew Bramble[1])

At Newington Green William was happy in the midst of what
he described to his mother as the 'unbounded love' of 'heavenly
minded friends', but it did not stop him dwelling in his letters home
on the 'vanity and shortness of life',[2] a reflection of the inadequacy
of eighteenth-century medicine and William's regular encounter with
death on the wards at St Thomas's. Nevertheless, his mother must have
been relieved to receive a letter from her brother, Richard Price, in which
he told her, 'I have much satisfaction in my nephew Billy.'[3]

It was too good to last. In July 1771, little more than a year after he
started at St Thomas's, William received the news of his sister Betsey's
death. She was twenty-six. When Dr Morgan wrote to William to break
the news, he was frank about the 'sorrow and tears' in the household. In
another letter, to a friend and patient, he wrote of his 'uncontrollable
affection' for his children and admitted, 'you cannot conceive how my
spirits have been depressed'.[4]

Only a few months later Jack, aged just fourteen, became ill with a
fever which quickly proved fatal. Dr Morgan was himself in poor health
by this time and having difficulty with the steep walk to their house.
The loss of another of his children depressed his spirits still further.
Dr Morgan died in the summer of 1772.

William had not completed his medical training but he knew what his father's wishes had been. He must return to Bridgend and take over the practice. He was desolate at the thought of leaving his uncle and admitted as much in a letter to his mother: 'The thought of separating from Dr Price damps every pleasure and checks the joy I should otherwise feel in returning home.' He was anxious, too, about being 'a young beginner in so small a place' and guessed that there would be local 'jealousies and ... adversaries'.[5]

William's misgivings proved all too accurate; his father's patients viewed him with deep suspicion. Not only was he young and inexperienced, worse, he was a cripple. His club foot was regarded as a weakness. How could anyone trust a doctor who could not heal his own malady?

Besides, there was by now a rival doctor in the town. Jenkin Williams, six feet tall, good looking, and quite the dandy in his gold-laced hat, inspired confidence in his patients. William didn't stand a chance. To make matters more awkward, he was to lose not only the practice, but also his sister, to his competitor, for Jenkin Williams was the suitor who had fallen for Kitty. Her misspelt letters had not deterred him, and in 1773 they were married. In the same year William relinquished the practice and returned to London.

His situation was bleak. He had not fulfilled the destiny his father had chosen for him, nor was he able to discharge the responsibilities he felt he owed to his widowed mother. He did what he could for her by renouncing his inheritance and making over to her the proceeds from the family estate at Tyle-coch, but the estate was small and the income was modest.

His only course of action was to seek out Richard Price and ask his advice. Price had a wide circle of influential friends and he had been a fellow of the Royal Society since 1765. He had published a number of papers on financial matters and was a regular consultant to the recently formed life assurance company, the Equitable Society, and this was to give William the opportunity for a new career. In the autumn of 1773, shortly after William's return to London, John Edwards, who held the key post of Actuary at the Equitable, died. A replacement was urgently needed. The exchange between William and his uncle has become part of actuarial folklore.[6] "'Billy,' Price said to his nephew, "do you know anything of mathematics?" "No, Uncle," the young man replied, "but I can learn."'

William set about learning mathematics, the very difficult mathematics of life assurance calculations. In April 1774, he was appointed Assistant Actuary at the Equitable on a salary of £100 per annum.[7]

Perhaps the events leading up to his appointment were not quite as fortuitous as they seem. John Edwards had been ill for some time, so his death could not have been entirely unexpected. Medicine had never been William's first choice; he had a keen mind and relished an intellectual challenge. If he recognised an opportunity and made certain that he was in the right place at the right time, who can blame him?

Not only did Price introduce his nephew to the Equitable, he stood surety for him. William had good reason to thank his uncle for his fresh start – and reason to thank another uncle. Samuel Price, half-brother to Richard Price, was a wealthy man and he also agreed to stand surety, pledging £500 'of good and lawful money'[8] should William prove unsatisfactory in his new post. William was hugely relieved at having secured his future. 'I am exceedingly obliged to you', he wrote to his uncle,

for this and all other favours conferred upon me, and I hope my future conduct will give you reason to believe that they are not altogether ill bestowed, as I know you want no other reward nor a more concerning proof of my gratitude than an upright and virtuous behaviour.[9]

He did exactly as he promised, and his hard work paid off. When, less than a year later, the death of the Actuary, John Pocock, meant that his post fell vacant, William was considered 'a very fit candidate' to be the next Actuary of the Society.[10] He was unopposed and unanimously elected by the directors.[11] Not yet twenty-five years of age, he was taking on a position of immense responsibility. It is inconceivable today that anyone so young and inexperienced could be appointed to such a post. A twenty-first-century actuary climbs the career ladder by means of study, training and examinations. William had invaluable guidance from Price and Price's publications[12] but in many respects he had to make it up as he went along. It is a measure of his success that he held the post of Actuary at the Equitable for the next fifty-five years, during which time the Society became one of the wealthiest corporations in the world.

As Actuary William was answerable to a Court of Directors, initially the group who had set up the Society – effectively partners who took responsibility for the Society's solvency. They met weekly; there was also the General Court which met monthly and where the policyholders could express their views (often quite stridently). William was in a sense everyone's servant. He did not always see it that way; as far as he was concerned he understood the Society's finances and *he knew best*. There were to be some fiery disagreements during his term of office.

William's first major undertaking as Actuary was to make a detailed examination of the Society's funds and to prepare estimates of the premiums and claims, tasks which took him from March to November 1775. When they were completed, he was asked to make a full valuation of the Society's liabilities. There were 992 contracts in force, each of which had to be separately valued. This took more than 2,000 calculations, but William managed to finish the valuation in just over four months. The columns of figures in his neat regular writing are a testament to his tenacity. They demonstrate hours of concentration, days and weeks of meticulous checking. These figures represent people's hopes and plans for their futures, their trust in the actuary's guardianship of their money. It is fitting that his hard work earned him a gratuity of one hundred guineas and the praise of the directors for his 'extraordinary diligence'.[13]

William's valuation confirmed that the Equitable's funds were very healthy – a state of affairs which needed to be considered in context. The nature of the business meant accumulation of capital. Most of this would be needed to pay the sums assured on the policyholders' deaths; this, after all, is the whole point of a life assurance policy. But, as William's calculations showed, there was some surplus, which arose for various reasons. For a start, loading was factored into the premiums as a precaution against an unprecedented cluster of claims. Then a surprising number of whole-life policies did not result in a claim, the policyholders having defaulted on paying premiums. Although the Society did have a policy of offering fair surrender values, very few policyholders took

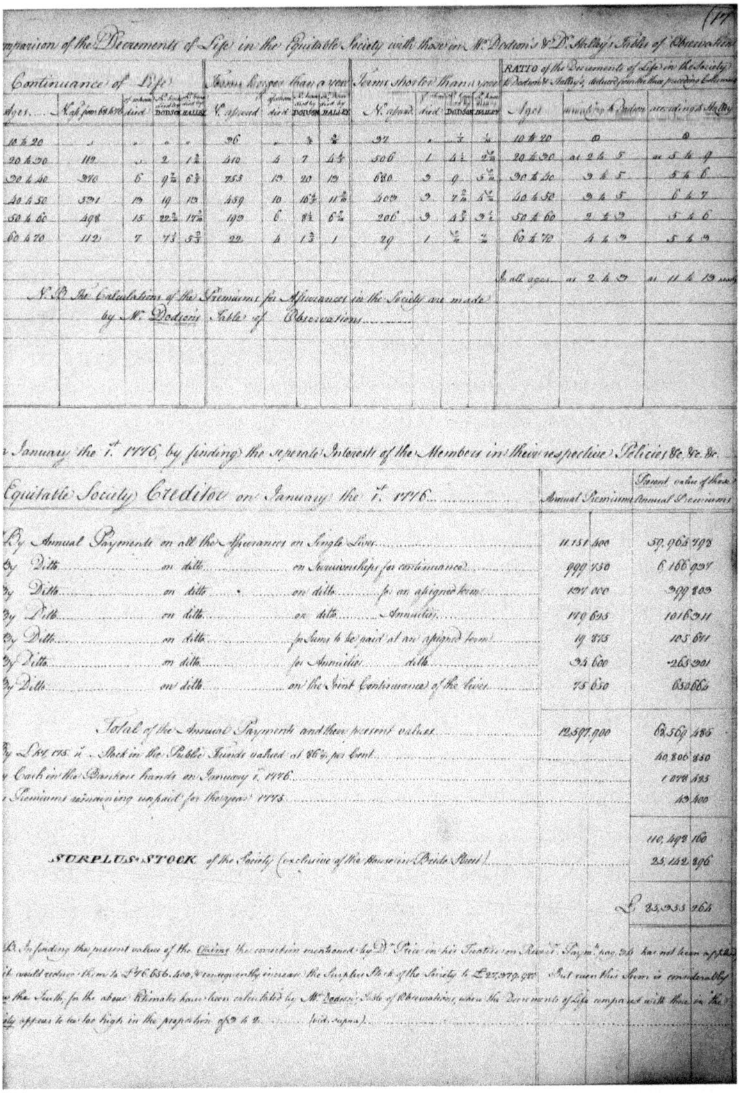

FIGURE 9 Society for Equitable Assurances on Lives and
Survivorships, William Morgan's 1776 calculations.
(Library Collection of the Institute and Faculty of Actuaries)

advantage of this and simply forfeited the premiums they had already paid.[14]

The Society invested funds in long-term government bonds which, for the purposes of calculating premiums, were assumed to yield a three per cent return. In fact, wars in Europe and America led to increased yields on money invested. As a result, as William's 1775 calculations showed, the Society had a surplus of some £25,000. This, in accordance with the Society's deed of settlement, might be divided amongst the members. It was an attractive option, but William agreed with Price's advice that they should retain the surplus 'as security against extraordinary events or a season of uncommon mortality'.[15]

Understandably (and sensibly) William leaned on Price in the early years of his appointment, but he made rapid progress. In 1779, only four years after becoming Actuary at the Equitable, William published *Doctrine of Annuities and Assurances*.[16] True, there is an introduction by Price explaining that he encouraged William to undertake the work, and that 'by his [nephew's] desire [he] revised it with some care'.[17] In addition, the book ends with an essay by Price on the state of the population in England and Wales. Clearly the publication was shaped as much by a Price agenda as a Morgan one, but Price praises William's efforts to 'render the subjects of which he treats as intelligible as possible to persons who may be unacquainted with mathematics'.[18] It's a generous assumption of what the non-mathematician can understand, for, after the opening paragraphs, the text becomes densely packed with figures and calculations. Nevertheless, the style throughout is clear and assured. William, only ten years after arriving in London as a penniless medical student, was at home in a different discipline and with difficult subject matter. He had come a long way in the intervening years.

As Actuary William was required to live at the Equitable Society's office – above the shop, as it were – so once again he had to take leave of his aunt and uncle at Newington Green. The Society's office was a house in Chatham Place on the corner with New Bridge Street. New Bridge Street – still there, if wider and busier now – was aptly named at the time for it led to the City from Blackfriars Bridge, which had been opened five years previously in November 1769. Before that there was

FIGURE 10 *A View of London taken from Albion Place, Blackfryars Bridge* (1802), a coloured aquatint by J. C. Stadler/N. R. Black.

Originally the bridge was a toll bridge, but the toll house was burned down during the Gordon Riots of 1780 and the tolls, always unpopular, were removed in 1785.

(Museum of London)

only London Bridge over the Thames although, outside the City, there were Putney Bridge, built in 1729, and Westminster Bridge, completed in 1750 despite opposition from watermen, ferry operators and the Corporation of London, who feared damage to City trade. Blackfriars Bridge restored the balance by creating a main artery into the City. For William, it meant that he was in the centre of a busy, modern area of the City. A contemporary painting shows official, commercial and domestic traffic crossing the bridge: soldiers in busbies, wagons carrying barrels, straw and logs, coaches and carts, as well as pedestrians.

The bridge was initially called William Pitt Bridge as a memorial to the popular prime minister,[19] but it became known, then as today, as Blackfriars Bridge. The Dominican Friars whose black cloaks gave the name to the area had departed in 1538 during Henry VIII's dissolution of the monasteries, but the priory buildings had remained and the old guesthouse had been acquired by the Society of Apothecaries. It was destroyed by the Great Fire of London but rebuilt, and is still the hall of the Worshipful Society of Apothecaries, one of the City of London's livery companies. The building, with its handsome entrance, acted as a daily reminder to William of the career he had abandoned. The portico entrance is today as it was in the eighteenth century; the Society's Coat of Arms surmounts the arched entrance to an inner courtyard and depicts Apollo, the god of healing, overcoming the dragon of disease, while above Apollo is the crest of the rhinoceros, whose powdered horn was thought to have wonderful healing powers. On each side a unicorn looks out to the world and below is the motto *Opiferque Per Orbem Dicor* (I am spoken of all over the world as one who brings help).[20]

William's medical training, although cut short, gave him know-how and contacts. He knew that controls were strict at the apothecary shops around the Society's hall, and were the best places to buy drugs of sound quality. Opium, for example, readily available and as widely used as paracetamol and aspirin today, could easily be adulterated, but William was well placed to know where to buy the best. When Samuel Price's widow, Catherine, became ill in her old age, William sent her 'large cargoes of opium' which he hoped would 'keep her well-fortified against the enemy'.[21]

FIGURE 11 The portico entrance to Apothecaries
Hall, with the Society's coat of arms above
the arched entrance to an inner courtyard.

Another benefit of William's medical training was some expertise
in judging the fitness of those wanting to take out life assurance policies
– a signed declaration of their state of health being a key part of the
process.[22] Applicants had to come in person to the Society's office after
11 o'clock on a Tuesday, where they met the Actuary and in his presence filled in the relevant forms, giving name, address, occupation and
age. Sometimes the assurance was to be on the life of another person,

in which case more details were required, not least to make sure that the policy was not merely a respectable front to gambling. In the early days of life assurance there was nothing to stop you effectively taking a bet on whether someone might die before a certain date, then collecting your winnings if the insured person lived beyond that date. In 1774, the Life Assurance Act put a stop to such 'gaming or wagering' by requiring that the insurer had to have a 'legitimate interest' in the person whose life was being insured.

Once the paperwork was completed, each applicant was ushered into the courtroom and the Actuary read the declaration. The directors could question the hopeful candidate before he (or quite often she) was asked to withdraw so they could discuss the proposal. The directors could agree to a whole-life policy, or they might restrict it to a few years, or even to only a single year; the decision had to be unanimous.

The list of those on whose lives policies were granted includes princes, dukes and other members of the aristocracy, as well as parliamentarians and members of the clergy.[23] These one might expect – also a host of literary names: Coleridge, Sheridan, Shelley, Southey, Byron and Scott. More surprising are the numerous 'ordinary' people, with a range of occupations: coachman, baker, carpenter, shoemaker, weaver, blacksmith, bookseller, cabinet-maker, and even one or two labourers. Some of the occupations speak of a distant way of life: pen-cutter, tallow chandler, Yeoman of the Revels, and even Table Decker to the Princess Dowager of Wales (something of a niche appointment in the royal household).[24]

For each of these new members William had to calculate the premium due according to their life expectancy and the length of the policy. He had much to do and he needed someone to manage his domestic affairs; he invited his younger sister, Nancy, to join him at Chatham Place as his housekeeper. Fortunately, as well as becoming commercially important, the locality had been generally improved in the 1760s and was no longer dangerous and unsavoury. The river Fleet, used as a sewer emptying into the Thames at Blackfriars, had been channelled underground, whilst site clearance for the construction of the bridge meant demolishing slums and cleansing an area previously filled with 'laystalls and bawdy houses, obscure pawnbrokers, gin-shops and alehouses; the haunts of strolling prostitutes, thieves and beggars'.[25]

Several Acts of Parliament in the 1770s and 1780s had done a lot to make the City and Westminster safer and healthier; shop signs, which – for all that they were attractive – made alleyways dark and hazardous as they were liable to fall, causing injury, were forbidden and some street lighting was introduced. As for the fashions of the day, London had much to excite twenty-three-year-old Nancy, who had so relished her season in Cowbridge. Just two miles east from Blackfriars was Cheapside, London's major shopping street with a wide range of luxury goods for sale. England was at peace after the Seven Years' War, trade with the rest of the world was expanding and – for the rich – the mid-1770s were boom years. The Swiss traveller César de Saussure, visiting London earlier in the century, marvelled at the shops which displayed 'the choicest merchandise from the four quarters of the globe' and where 'a stranger might spend whole days, without ever feeling bored, examining these wonderful goods'.[26] Writing only fifteen years after Nancy's arrival in London, the German novelist Sophie van la Roche described the delights of 'lovely Oxford Street . . . a street taking half an hour to cover from end to end . . . First one passes a watchmaker's, then a silk or fan store, now a silversmith's, a china or glass shop'.[27]

There were, as well, confectioners and fruiterers with voluptuous displays and spicy scents of exotic produce: 'pyramids of pineapple, figs, grapes, oranges and all manner of fruits'.[28] Even the booths selling cheap gin and other drinks looked attractive, with lights behind the crystal flasks making the different-coloured spirits sparkle.

There was, however, a harsher side to the London streets. The post of reformation (the stocks) in Cheapside was still used as a place of punishment for non-marital sex or sodomy. Public executions still took place at Tyburn and attracted large numbers of spectators. Contemporary pictures show that, even on festive occasions such as the Lord Mayor's Procession, crowd behaviour was often rowdy, drunken and disorderly, a rich field for pickpockets and prostitutes. Sometimes the crowds gathered to protest; disturbances were not uncommon in the eighteenth century, bread and food riots being particularly frequent.

Benjamin Franklin, writing in 1769, produced a long list: 'I have seen, within a year, riots of colliers, riots of weavers, riots of coal-heavers, riots of sawyers, riots of Wilkesites, riots of government chairmen, riots of smugglers.'[29]

But it was anti-Catholicism which sparked the riots in London in the summer of 1780 – six days of death and destruction which were a direct threat to William and Nancy at Chatham Place.

4

1780 – FLAMING JUNE

Who can but for a moment think on the danger, without looking up to heaven in grateful acknowledgment to the Supreme Being for so signal a national deliverance?
(*The Gentleman's Magazine*[1])

Religious differences continued to divide the nation and many regarded Catholics with dislike and suspicion. The plaque blaming the Fire of London on 'barbarous Papists' had been reinstated on the Monument in 1689 and remained as a daily reminder to Londoners of potential conspirators in their midst. The Test and Corporation Act imposed constraints on the civil rights and religious practices of Roman Catholics but, in 1778, the Catholic Relief Act removed some of these restrictions, most notably the requirement that those who enlisted in the British Army had to condemn the Catholic Church as part of the oath of allegiance to the Crown. This concession, making it easier to recruit Catholics, was a barely disguised design to boost the flagging numbers of troops fighting in the colonies – the American War of Independence. The Catholic Relief Act passed through parliament without incident but things were very different when attempts were made to extend the law to Scotland, and rioting broke out in Edinburgh and Glasgow. Unrest, championed by Lord George Gordon, MP for Inverness-shire, spread south, and a freshly invigorated Protestant Association gathered some 44,000 signatures on a petition for the repeal of the Catholic Relief Act. The day selected to present the petition to parliament, Friday 2 June 1780, was hot and sultry, the sort of short-fuse day when frayed tempers might be expected to ignite into angry actions. Which is what happened. There followed six days of rioting which became known as the Gordon Riots. The violence of the mob terrified the residents of London and, at Chatham Place, William and Nancy were in the thick of it.

Events began with the protesters (estimated to number 60,000) gathering at St George's Field, halfway between Westminster and Blackfriars Bridge on the south side of the Thames.[2] Contemporary accounts record an orderly crowd who 'made a noble appearance and marched in a very peaceable and quiet manner'. After an address by Gordon they divided into three sections, each taking a different bridge across the Thames. The second section crossed by means of Blackfriars Bridge, so William and Nancy at Chatham Place were in the direct path of some 20,000 protesters as the column of men, blue cockades in their hats, advanced across the bridge waving flags and chanting hymns whilst, in their midst, Scotsmen skirled their bagpipes.

By the time the marchers reached the Houses of Parliament the mood had become ugly; MPs were roughly handled and their carriages damaged. The Lord Chief Justice and the Archbishops of York and Canterbury all had their wigs torn off, and even the Prime Minister, Lord North, suffered at the hands of the mob. His carriage was surrounded and the door opened. The Horse Guards arrived in time to save him from being dragged into the street, but not before his hat had been seized by an enterprising rioter who later cut it into pieces, each of which he sold for a shilling[3] – enough in 1780 to dine out on beef, bread and beer in a steakhouse.[4]

The rioters became more violent, spreading mayhem throughout London, and targeting not only Catholic chapels, but the houses of judges, politicians and the rich. On Tuesday 6 June, first Newgate and then Bridewell, the Fleet and the King's Bench prisons were stormed and, once their prisoners had been released, burned. These – in particular Newgate, which had been recently enlarged and a new sessions house built – represented an attack on symbols of authority and indicated unrest at a deeper level, even the stirrings of revolution. Eyewitness accounts in newspapers and in private letters paint a terrifying picture of flames so high that 'the sky was like blood with the reflection of them', whilst the mob could be heard knocking the irons off the prisoners, together with 'the shouts of those they had released, the huzzas of the rioters, and the universal confusion of the whole neighbourhood'.[5] Within weeks, the print shops were selling images of the destruction of Newgate. They depict Newgate in flames as jubilant crowds dance in celebration beneath a flag declaring 'No Popery'.

Arguably, the cartoons sensationalise the rioting, but these are events which indisputably took place.[6] Given the similarities with the French Revolution at the end of the decade, it is surprising that the Gordon Riots get only brief coverage in general history books. In fact there was considerably more damage done to property, both public and private, than in Paris in 1789, and yet, outside academic circles, the Gordon Riots have faded from the collective memory – just occasionally jogged by works of literature. They are the setting for *Barnaby Rudge*; Dickens gives imagined descriptions of mob violence throughout the novel, culminating in the destruction of Newgate: 'the deafening tumult' overlaid with the 'clash of iron ringing upon iron' as the rioters struck

FIGURE 12 *An Exact Representation of the Burning, Plundering and Destruction of Newgate by the Rioters, on the memorable 7th of June 1780* (pub. 1781). *(British Museum Satires 5844, hereafter BM Satires)*

the prison wall with sledgehammers; heat so intense that the paint on nearby buildings swelled 'into boils, as it were from excess of torture' as the roofs collapsed, and 'sparrows rendered giddy by the smoke, fell fluttering down upon the blazing pile'.[7]

Nancy never forgot the riots, and when, the following year, she returned to Bridgend, she regaled her family and friends with the story of her encounter with the angry mob. For William the riots were not only a danger to his and Nancy's personal safety, they were a threat to the documents at Chatham Place.[8] At Blackfriars the tollbooths on the bridge were destroyed whilst, less than a mile away, the Bank of England was targeted, a deployment of the troops called in to defend it, and rioters killed in the gunfire. So serious was the attack and so dire the possible consequences that, even when the riots were quelled, the Bank was given nightly protection by soldiers – a practice which continued until 1973.

Official figures put the dead at 300, but the number was undoubtedly much more, some eyewitnesses putting it as high as 700.[9] Afterwards fifty-six people were sentenced to death, but there were so many pleas for pardon that eventually only twenty-six were hanged. Their executions did not, as might have been expected, take place at Tyburn – a measure of the anxiety that rioting might erupt again. Tyburn was notorious for the numbers attending and for the grotesque carnival atmosphere of noise, drunkenness and disorder. Instead, gallows were erected at different places, near to the particular scene of crime; spectators were fewer in number and quite peaceable.[10]

As to the guilt of those hanged, the list of their names, ages and occupations argues that their misfortune was to have been caught up in the rioting, their crimes opportunistic rather than motivated by politics or religion.[11] One contemporary recollection comes from Samuel Rogers, who remembered seeing a cartload of girls dressed in coloured dresses passing through the streets of the City on their way to execution. They had been condemned to death for taking part in the riots though, in Rogers's view, they had probably done little more than look on.[12]

The underlying causes of the Gordon Riots remain open to debate. At the very least, it was naive of Lord George Gordon not to guess that the marchers would become restless and angry. As the lawyer Samuel Romilly commented at the time: 'What! – summon 40,000 fanatics to meet together, and expect them to be orderly? What is it but to invite hungry wretches to a banquet, and at the same time to enjoin them not to eat?'[13]

Lord George Gordon was tried for high treason but acquitted, on the grounds that he had not planned – or even foreseen – the riot. What he organised was only a march to present a petition to parliament. His acquittal owed much to his junior defence lawyer, Thomas Erskine, who, at the age of thirty-one, had recently been called to the bar and was beginning to make a name for himself – a name which was to reappear in future landmark trials, in particular the 1794 treason trial which William was subpoenaed to attend.[14]

Politically the rioting can be seen as a measure of rumbling dissatisfaction with the government and the economic situation – and the war with America. Britain had by this time been at war with America for five years; a war, albeit one being fought more than 3,000 miles away, which was ever present in conversation and correspondence at Newington Green.

5

AT WAR

I wish as ardently as you can do for Peace and should
rejoice accordingly in cooperating with you to that end.
(Benjamin Franklin[1])

At Newington Green William and Nancy were frequent visitors,
as was their brother George, who by 1773 was also in London.
George had, as planned, gone up to Jesus College in 1771, but he left
after only one year. Possibly his father's death meant that his mother
could no longer afford to support him at Oxford, but it seems more
likely that there had been a change in his religious views, because he
enrolled at the Hoxton Dissenting Academy.

The move is revealing and indicates that George, probably influenced by his uncle, Richard Price, was by this time moving towards
Unitarianism.[2] The terms of the detested Test and Corporation Act
remained in place, despite the Toleration Act of 1689, which had not
been as accommodating as the name implies. Nonconformists such as
Baptists and Congregationalists were granted freedom of worship, but
the Act specifically excluded Unitarians and Roman Catholics (also
Jews and atheists) from political office – and from studying at Oxford
and Cambridge. A change in his religious outlook would have made it
impossible for George to remain at Oxford.

From an academic point of view the move was greatly to his advantage. At the end of the eighteenth century Oxford and Cambridge were
better known for drunkenness and debauchery than for scholarship and
study. Jonathan Swift wrote of 'the idleness and the drinking' at Oxford,

and many of his contemporaries complained that the universities were 'neglectful and inefficient in the performance of their proper work'.[3] The Dissenting academies, on the other hand, were effectively alternative universities, and provided a rigorous education to a high standard and with a wide curriculum: logic and metaphysics, Jewish and Christian antiquities, ethics, scripture, mathematics, science, modern languages and history.

George continued his studies in the Classics but, at Hoxton, he became seriously interested in mathematics and science – subjects which he was to continue studying throughout his life. During his summer vacation in 1773 he assisted Price, who was drawing up tables of life assurance to present to parliament.[4] At Newington Green there were, of course, the telescope, the microscope and the electrical machine so much enjoyed by the Rogers children. As well as his lifelong interest in electricity, George was fascinated by the night sky viewed through the telescope.[5]

Just as exciting for William, George and Nancy was the intellectual atmosphere at Newington Green, for many years a place which had attracted and nourished radical thinking. Several Dissenting academies had flourished in the area; alumni included Daniel Defoe (1660–1731) and the hymn writer Isaac Watts (1674–1748). With Price's arrival Newington Green became an 'enlightenment hothouse';[6] the list of those who visited, or who corresponded with Price, reads like an eighteenth-century *Who's Who* and includes philosophers, economists and political activists – David Hume, Adam Smith, John Horne Tooke, Tom Paine, Mary Wollstonecraft – as well as the prison reformer John Howard, American Founding Father Thomas Jefferson, American President John Adams, and the polymaths Joseph Priestley and Benjamin Franklin.[7] Joseph Priestley, Nonconformist priest, political radical and brilliant scientist, was a friend of long standing and, by 1773, he was working as librarian to the Earl of Shelburne, an appointment which gave him time to conduct experiments.[8] Scientists were by this time daring to question the long-held belief that the basic ingredients of the material world were Earth, Water, Fire and Air.[9] Obvious today, but at the time a challenge to accepted ideas quite as revolutionary as the Copernican theory that the earth is not the centre of the universe. Priestley's experiments led to an understanding that the supposed elements are in fact compounds, as well as to his famous discovery of oxygen, which he called dephlogisticated air.

William later described him as 'an admirable philosopher' who made 'astonishing discoveries', while Franklin was one of 'the brightest ornaments' of coffee house society.[10] 'Ornament' is still used to denote a person who adds distinction to his sphere and time,[11] but to a modern reader it is a modest compliment to someone whose achievements were so important and wide-ranging. After his early beginnings as a writer and printer, Franklin became a scientist, inventor, postmaster, political activist, Freemason and, eventually, a diplomat and one of the American Founding Fathers.

A polymath with insatiable curiosity and an inexhaustible range of interests, his passionate enthusiasm must have been inspirational. He declared himself 'almost sorry [he] was born so soon' since it meant he could not have 'the happiness of knowing what will be known 100 years hence'.[12] His inventions included bifocal spectacles, a urinary catheter, a musical instrument which he called the 'armonica', and the Franklin stove, whose heat went into the room rather than up the chimney. Most celebrated was – and is – his invention of the lightning rod and the understanding of electricity which underlies it. Pictures of Franklin's experiment with a kite in a thunderstorm have become iconic, even appearing on American postage stamps. Franklin claimed that he deliberately flew a kite in a thunderstorm, with a key attached by a silk string to the kite. When the lightning struck a charge ran down the wet silk and into a Leyden jar; Franklin, safe and dry in a nearby shed, was thus able to demonstrate that lightning is static electricity. It's an experiment not to be recommended for trying at home since it can lead to electrocution, but, whilst the truth of Franklin's story has been challenged,[13] similar experiments were successfully carried out in the second half of the eighteenth century, and many of Franklin's contemporaries were warm in their congratulations. Kant hailed him as a 'modern Prometheus';[14] Priestley wrote of Franklin's 'great and deserved reputation';[15] and, in 1753, the fellows of the Royal Society bestowed on him their highest honour – the Copley Medal – 'on account of his curious Experiments and Observations on Electricity'.[16]

William's brother, George, was moved to even more extravagant praise: Franklin had 'wrenched the thunderbolt from the grasp of tyranny and fraud'.[17] George saw science as releasing us from the oppression caused by superstition and its 'ruling priesthood'. When we understand thunder and lightning, we no longer fear it as a demonstration of wrath from God. Hitherto it had been common practice, especially on the continent, to ring cathedral bells during a thunderstorm to protect the community from lightning strikes.[18] 'It is a dogma of faith', wrote Thomas Aquinas, 'that the demons can produce wind, storms, and rain of fire from heaven … [but] the tones of the consecrated metal repel the demons and arrest storms and lightnings'.[19] When lightning strikes were avoided, Aquinas's theory conveniently appeared proven; when lightning killed bell-ringers, their deaths were ascribed to God's judgement for sins.

Franklin lived for nearly sixteen years in London, but by the summer of 1775 Britain was at war with her American colonies and, as a colonist, he had to return to America. The situation in America became Price's chief concern during the 1770s and he was in regular correspondence with Thomas Jefferson and John Adams, as well as Franklin. In addition, he made his arguments known to a wider public through his published pamphlets and sermons.

William agreed with his uncle's views and admired his courage in expressing them. In his *Memoirs of Richard Price* he uses language that leaves no doubt about his own opinions. Taxing the Americans to repair Britain's 'dilapidated finances' was a 'direct attack on the civil liberties of the American colonies', which would plunge the nation into a war to 'enforce unconditional submission to claims which were manifestly unjust' – a war which was an 'unnatural contest'.[20]

As for the Declaratory Act of 1766 which gave the king and parliament full power to make laws binding on the colonies, this was a 'tyranny … expressed in its strongest terms', and William emphasises his outrage by using italics to quote the offending resolution in full. He labels the legislation of subsequent years as alternating acts of 'violence and concession', culminating in the 1770 Act which repealed the Townshend Duties on every article except – and here William returns to italics – '*a duty of one penny a pound on tea*'.

Taxation without representation – this was the cry that led to the Boston Tea Party. The East India Company, whose fortunes had been hit by a decline in tea sales coupled with the rising cost of policing its commercial empire, needed help. In return for a loan from the British government, the Company was given leave to export tea directly to the American colonies, thereby raising tax revenue for Britain. When large quantities of tea were shipped to America it imposed a tax burden which infuriated the colonists, and in Boston the tea was thrown into the sea. The British response was to close the port for trade and to send a detachment of soldiers to America. Effectively war became inevitable.

William felt that 'a great many of the more enlightened and virtuous' not only deplored the measures Britain was taking, but believed they threatened the 'rights and liberties' of the British as well as Americans.[21] Events at home supported his claim. When news of the battle of Lexington reached England – a battle begun by 700 British soldiers firing three volleys at a little troop of seventy men – Samuel Rogers's father, Thomas, put on mourning. Asked if he had lost a friend, he replied that he had lost several friends – in New England.[22] He was not alone. The Recorder of London also dressed in mourning for the same reason.

Price, meanwhile, spent the winter of 1775 working on a pamphlet, *Observations on the Nature of Civil Liberty, the Principles of Government, and the Justice and Policy of the War with America*. It was published in February 1776 and the first edition of 1,000 copies sold out within three days. Further editions sold quickly, and supporters of the Americans felt that their cause would be best served if the pamphlet could be made available to what William called 'all ranks of society'.[23] Price agreed to a cheap edition, thus sacrificing the 'very considerable' sum he would have made from the 60,000 copies that were sold in the next few months. In addition to the English editions, the pamphlet was translated into French, German and Dutch, and there were reprints in America.[24]

'Dr Price's name was in everyone's mouth' and, as he rode through the streets of London on his old white horse, 'he was often diverted by hearing the carmen and orange-women say, "There goes Dr Price! Make way for Dr Price!"[25] It was not just the man and woman in the street who admired his views. In March 1776 the Council of the City of London conferred on Richard Price the freedom of the City of London, which was presented to him in a gold box 'as a testimony of [their] approbation for his late pamphlet'.[26]

Not everyone agreed with his views or approved of his publishing them. William is dismissive as he lists Price's critics. Edmund Burke is 'that very equivocal friend of liberty';[27] Burke had been sympathetic to the American complaints but, as William points out, his concern was not with the morality of taxing the colonists but with the feasibility of doing so. Amongst the clergy Price's adversaries were 'in dreadful array' and included the Archbishop of York and John Wesley, all denouncing 'their anathemas against the friend of conciliation and harmony'. Then there were the pamphleteers, many of them in the pay of the government, who wrote in answer, or rather 'in abuse', of Price's arguments.

As hostilities intensified Price received anonymous letters threatening his life. These he concealed from his wife, but she was aware that their house might be searched, and she took packets of his papers next door to Thomas Rogers for safekeeping.[28] William became a conduit for his uncle's mail. In several letters Price asks his correspondents to address their letters to 'Mr William Morgan at the office for Equitable Assurances near Black-Fryars Bridge London', adding in one 'they will in this way be less in danger of miscarrying'.[29]

In 1778 France, never slow to renew hostilities with a long-time enemy, declared war on Britain, followed in 1779 by Spain. In addition, opposition to Britain came from the League of Armed Neutrality founded by Sweden, Denmark and Russia, then joined by Holland, Prussia and Portugal. The League was formed to resist the British blockade of the colonies and her demand to search neutral vessels; one consequence was Britain's declaration in 1780 of war with Holland for supplying arms to the colonists.

British forces were stretched on land and sea. In 1781 Lord Cornwallis in Yorktown on Chesapeake Bay had fewer than

9,000 men in the face of some 16,000 French and American troops. He surrendered. When, on Sunday 25 November, the Prime Minister, Lord North, received the news of the defeat of Yorktown he is reported to have reacted 'as he would have taken a ball in his breast . . . he opened his arms, exclaiming wildly, as he paced up and down the apartment during a few minutes, "O God! it is all over!"'[30]

The peace treaty was eventually signed at Versailles little more than a year later, in January 1783. Britain ceded Florida and Minorca (but not Gibraltar) to Spain; France gained Senegal and Tobago; and Britain recognised the independence of America. A momentous event, but cut to size by a cartoon which appeared a few weeks later.[31] Its title *The General P—s or Peace*, it shows five men, representing Britain, Holland, France, Spain and (in the person of a Native American) America – the main participants in the war. They have laid aside their swords, flags and drums and are urinating into a communal pot. A speech-bubble from the Frenchman boasts, 'We have wrangled you out of America'; one from the American calls it 'a free and Independent P—', while the Englishman protests, 'I call this an honourable P—'. Colourful if crude, the image points a cynical comment underlined by a rhyme beneath the picture:

> Come all who love friendship, and wonder and see,
> The belligerent powers, like good neighbours agree,
> A little time past Sirs, who would have thought this,
> That they'd so soon come to a general P—?
>
> The wise politicians who differ in thought,
> Will fret at this friendship, and call it to nought,
> And blades that love war will be storming at this,
> But storm as they will, it's a general P—.
>
> A hundred hard millions in war we have spent,
> And America lost by all patriots consent,
> Yet let us be quiet, nor any one hiss,
> But rejoice at this hearty and general P—.

Tis vain for to fret or growl at our lot,
You see they're determined to fill us a pot,
So now my brave Britons excuse me in this,
That I for a Peace am obliged to write Piss.

FIGURE 13 *The General P——s, or Peace*, unattributed
cartoon (J. Barrow, Blackfriars Bridge, 1783).
(Library of Congress)

A BEAUTIFUL GREEN LIGHT

Is electricity a subtile elastic fluid? or are electrical effects merely
the exhibition of the attractive powers of the particles of bodies?
Are heat and light elements of electricity, or merely the effects
of its action? Is magnetism identical with electricity, or an
independent agent, put into motion or activity by electricity?

(Sir Humphry Davy[1])

The 1780s began with a discovery that thrilled the scientific world: a
new planet. William Herschel, a Hanoverian musician, had moved
to Bath in 1766 to become organist and choirmaster at the Octagon
Chapel.[2] As well as composing, teaching the guitar, harpsichord and
violin, and giving singing lessons, Herschel found time to study the
night sky. Joined by his sister, Caroline, in 1772, he began to design and
make his own telescopes. The resulting instruments were more accurate
and more powerful than any he could buy, and what they revealed was
truly amazing.

Not only did Herschel discover Uranus, the seventh planet in the
solar system, but he and his sister catalogued comets and stars numerous
beyond previous speculation. Suddenly the cosmos was immeasurably
vast and unimaginably old. For some these new ideas challenged biblical certainties and began a process of doubt that would re-emerge in
the next century.

Richard Price was excited by the new ideas[3] and, had William's
diary survived, it would surely have included his thoughts on Herschel's
discoveries. Perhaps, too, he recorded his views on the news in 1783

from France of the first manned flight in a hot-air balloon. Very likely he shared Price's view that 'the discovery of air balloons seems to make the present time a new Epoch'.[4]

In fact William was also contributing to a 'new epoch', though it would be years before the value of his electrical experiments was recognised. There was a continuing debate about the nature of electricity. Clearly electricity moved – or flowed – from one body to another (as in Abbé Nollet's line of soldiers all jumping from one electric shock) and, if electricity flowed, was it a fluid? Franklin thought so. In his view all substances in their 'unchanged condition' possess a fixed quantity of electric fluid. He also thought that there were two forces at work, and he coined the terms positive and negative; here he was on the way to recognising that there are two *types* of force.

If, demonstrably, electricity moved through objects and through air, what about a vacuum? William set out to investigate the question, and in 1785 Price presented a paper on his nephew's behalf to the Royal Society: 'Electrical Experiments Made in Order to Ascertain the Non-Conducting Power of a Perfect Vacuum'.[5] William's account is clear and detailed; it compares well with modern standards in experimental reporting. He makes it clear, as well, that he had witnesses to his experiment, amongst them his uncle, Richard Price, and a Mr Lane (probably Timothy Lane FRS (1734–1807), who described himself as 'apothecary and electrician' and was later to support William's election as a fellow of the Royal Society).

William's apparatus consisted of a 'gage', a tube about fifteen inches long, inverted through a brass lid into a small vessel containing mercury. The top of this gauge was wrapped in tinfoil (a good conductor) and a small wire (another good conductor) fixed to the cap of the vessel reached down into the mercury. The pressure difference between the inside and the outside of the inverted tube would have created a vacuum – evident as the height of the mercury dropped. The vacuum made by this technique, however, is not perfect because air will normally be dissolved in the mercury and slowly escape into the vacuum. So, before inverting the gauge, William boiled the mercury in order to release the trapped air, which could then be pumped away.

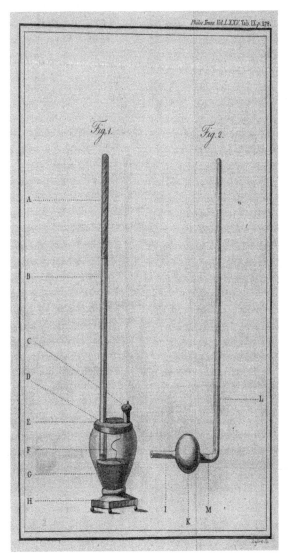

FIGURE 14 The apparatus for William Morgan's 'Experiments Made in Order to Ascertain the Non-Conductive Power of a Perfect Vacuum', reported in *Philosophical Transactions of the Royal Society* (1785). *(The Royal Society)*

The experiment was a hazardous one. Today mercury is considered by the World Health Organization as one of the top ten chemicals of major public concern – with good reason for, in its vapour form, mercury is known to have toxic effects on the nervous, digestive and immune systems and on lungs, kidney, skin and eyes. The expression 'as mad as a hatter' comes from the fact that hatters, who used mercury in hat-making, notoriously suffered from mental disorders. Lewis Carroll tapped into the idea for his mad tea-party in *Alice's Adventures in Wonderland*.

If William was aware of the danger of his experiment perhaps he saw it as just another risk to assess and not so very different from the actuarial calculations of his daily work. What he does admit is that the procedure required 'some nicety and no inconsiderable degree of labour and patience' (277) – something of an understatement, since just boiling the mercury took three to four hours. Creating a vacuum was a tricky business and he did not always succeed. He does not tell us how many times he repeated the procedure before he was certain of the result, and recorded that 'neither the smallest ray of light, nor the slightest charge, could ever be procured in the exhausted gage' (273). He had demonstrated that a perfect vacuum was a non-conductor.

It is his failed attempts that are of scientific and historical importance. He noted that 'if the mercury in the gage be imperfectly boiled the experiment will not succeed; but the colour of the electric light which, in an air rarefied by an exhauster, is always violet or purple, appears in this case of a beautiful green'.[6] 'A beautiful green light' has an appropriately symbolic ring; it certainly was symbolic – and far more. The green light William observed has led to the claim that he produced the very first X-ray tube. True he did not know that he had done so, nor was it an X-ray in the form that we know it today, but he did notice something which he considered 'very curious': the colour of the light in the tube showed how near – or far – he was from creating a vacuum or, as William put it, indicated 'the degree of the air's rarefaction'.

It was more than a hundred years later that Wilhelm Conrad Röntgen, in 1895, discovered the power of X-ray to 'see' through flesh and blood, terrifying his wife, Bertha, by showing her, by means of

X-ray, the bones of her hand. Röntgen used X, the mathematical term for an unknown quantity, because he did not understand the nature of the rays he'd discovered; the name stuck. A century later, in a 2009 poll marking the Science Museum's centenary year, members of the public voted the X-ray machine as the scientific advance with the greatest impact. William could not grasp the world-changing potential of his green light, but his meticulous experiment was a starting point in the journey that led to Röntgen's discovery.

So what did William do? For a start, he created a vacuum – no mean feat with the very basic tools at his disposal. As for when 'the mercury was imperfectly boiled', he then had a partial vacuum. What he observed in that partial vacuum was the conversion of gas to plasma. Plasma is one of the four fundamental states of matter, the others being solid, liquid and gas. The passage from a gas state to a plasma state is ionisation, and happens when a gas is exposed to a very high temperature or – as in William's experiment – when a gas exists where there is a large voltage difference between two points. Scientists today understand that the conversion of a gas to a plasma causes electrons to leave the atoms and so they increase the number of free electrons. Decreasing or increasing the number of electrons in atoms of the gas creates positive or negative charged particles called ions. What William observed was the ionisation of a gas: the gas within his partial vacuum was split into ions which bombarded the glass tube and produced fluorescence – and soft X-rays.

William's experiment would be as laborious and as hazardous for us to repeat today as it was for him. But we do not have to recreate the whole procedure to see what he did. Plasma balls – those glass globes with streaks of coloured light which dance in fantastical patterns – are made by ionising gas within the sealed glass sphere. They have an electrode at the centre that is at a very high voltage, and there is a low pressure of inert gas, such as argon, in the ball. The high voltage causes the gas to ionise and a current flows through it producing light. If you touch the plasma ball the coloured rays change into new configurations. That is because your finger acts as a second electrode; the current flows from the central electrode, through the gas, then through the wall of the ball and through your body to the earth. The current is small so you are not in any danger.

But William's partial vacuum was a lot more than the forerunner of a novelty toy. It was effectively the first link in a chain of investigation. In December 1821, the year of his election as President of the Royal Society, Sir Humphry Davy presented a paper[7] in which he cited William's experiment and took it further. Pointing out that, even in the best vacuum which could be made with William's equipment, there would still be some vapour of mercury, 'though of extremely small density',[8] he questioned the accuracy of the 1785 experiment and, following his own experiments, came to a conclusion in direct opposition to William's: that a vacuum (as perfect as could be made) *could* conduct electricity.

William would have met Davy at the Royal Society as well as, most probably, socially through Samuel Rogers and Horne Tooke, but, if the two men discussed their respective experiments, no record remains. In 1821 William was preoccupied with concerns about the Equitable and the national economy; he was silent on his feelings about Davy's paper. He did not live long enough to learn of Michael Faraday's paper[9] in 1838, in which he challenged Davy's conclusions and reported that, in his experiments, he observed that the luminous discharge was principally on the inner surface of the glass. He thought, therefore, 'it does not appear unlikely that, if the vacuum refused to conduct, still the surface of glass next [to] it might carry on that action'. Faraday was on to something. In 1875 William Crookes invented the Crookes tube – a partially evacuated glass bulb in which ionisation could take place less laboriously than in William's experiment. Röntgen used a Crookes tube for his momentous discovery in 1895 – 120 years after William's 'beautiful green light'.

RISK AND REWARD

Those who aspire not to guess and divine but to discover
and know; who propose not to devise mimic and fabulous
worlds of their own, but to examine and dissect the nature
of the world itself must go to facts for everything.

(Sir Francis Bacon, *Instauratio Magna* (1620)[1])

Electrical experiments were William's hobby. At work he was busy
with the steady stream of applicants for life assurance. The 1780s saw
policies issued to cover the lives of the Prince of Brunswick, the Dukes
of Cumberland, Chandos and Devonshire, the Earls of Clanrickarde,
Shelburne, Derby and Donegal, and Lord Melbourne.[2] There were also policies for 'gentlemen', for merchants and grocers, for coachmen and labourers
and even one for Mary Wells, a comedian at Covent Garden Theatre.[3]

A number were for young men in the military, the terms of their
policies saying much about the risks their lives involved as well as suggesting a story of loans and debts behind each policy.[4] Why, for example,
on 29 January 1783 did Robert McLeroth of Belfast insure the life of
Maurice Keatinge, aged twenty-two, a captain in the 22nd Regiment
of the Dragoons, for £1,600 and for a period of only three calendar
months?[5] Not only did he pay a premium of £8 4s. 6d and an entrance
fee of £10, but on 5 February 1783 he paid an additional £8 so that
the policy 'be not vacated by the said Captain Keatinge's dying upon the
seas in his passage from Great Britain to Ireland'.

William had to calculate the premiums. For the policyholder in an
insurance pact it is a gamble: none of us knows when we are going to

die. An actuary, if he has got reliable data and if he has got his sums right, can tell us how long we might expect to live.[6] But in the mid-eighteenth century there were no proven rules for 'getting it right', and at least ten societies had sprung up in the 1760s only to fold after a few years, amongst them the Law Society for the Benefit of Widows,[7] the Provident Society, the Rational Society and the Friendly Society of Annuitants.[8] Their failure was in part due to their use of inadequate statistics with which to estimate life expectancy. Basic data about births and deaths was, before the establishment of regular censuses, patchy and unreliable. What records there were – for example, baptismal and tax records – were not uniform or centralised and were often incomplete.[9] (The absence of any baptismal record for William himself is an ironic reminder of the problems involved!)

An early example of a systematic analysis of data was the table of deaths for the German city of Breslau (now Wroclaw in Poland) created by Edmund Halley (1656–1742).[10] Halley's methodology was, in fact, key to the creation of actuarial science, but his sample material was limited and skewed by his exclusion of Catholics. Another table, created by John Smart, of deaths in London between 1728 and 1738 was flawed by its covering only a nine-year period and one when cheap gin had inflated the figures.

William had an important advantage. Richard Price had constructed a table showing the mortality records made over a forty-five-year period for the parish of All Saints in Northampton and checked against a local census. Given that the ages of the dead were 'normal' and not distorted by exceptional circumstances, they provided a guide to the lottery of life expectancy that was more accurate than the previous attempts. The data came to be called the Northampton Table and played a hugely significant role in the early development of actuarial science. After 1782 and throughout the rest of his career, William used this more reliable table, initially adding fifteen per cent to the premiums as a precautionary 'loading' in the Equitable's favour. (His reluctance to recognise that there were imperfections in the Northampton Table was, however, to cause him problems in later years when up-and-coming actuaries challenged him.[11])

With an individual policy, however, the element of gamble remained (and still does). In the case of an annuity, if the odds are on a life lasting seventy years, the man or woman who makes it to eighty or ninety has done very well out of the bargain. If he or she falls under a bus – or a stagecoach – the day before drawing a pension, the insurance company is the winner. If the insurance is for the benefit of a widow, her husband will be seeking to cushion her financially after his death. In the event of an early death the insurance company is the loser, having received only a few of the premiums due on the policy. In each case, although the situation is different, the element of chance is similar. It is up to the actuary to work out the premiums and payments based on the balance of probabilities. Central to the success of the actuary is the mathematics of probability and especially the mathematics of large numbers. The policyholder gambles a single bet (on his or her length of life), while the actuary is taking as many bets as there are policyholders in the life assurance company. Actuarial science 'works' because the mathematical laws of large numbers enable the resulting average age of death of the members of the group to be accurately calculated.

Once again Price had done much to prepare the ground for William. In the first place there was a paper, published in 1763 by the Royal Society, entitled 'An Essay towards Solving a Problem in the Doctrine of Chance'.[12] This was an edited and revised version of a mathematical essay left in the papers of a friend, Thomas Bayes, who had died some three years previously. It seems likely that the theorem Price presented was as much his own work as Bayes's, and that without his revisions it might even have sunk without trace, but Bayes gets the credit. Known as Bayes's Theorem, it is now hailed as a 'milestone in the history of mathematics'.[13] It is used today in a range of fields, from modelling climate change to astrophysics, from stock-market analysis to filtering email spam.[14] Put simply, it is a way of reaching statistical probabilities based on partial information.[15]

Two more papers from Price had followed: in 1769 'Observations on the Expectations of Lives'[16] and then 'Observations on Reversionary Payments', which was published in 1771 and went through seven editions, the final three of which were edited by William after Price's death.[17] *Observations on Reversionary Payments* became essential reading for anyone involved in actuarial science and remained the pre-eminent work well into the nineteenth century.

Building upon this foundation, William's own work took the mathematical theory further with two papers which Richard Price presented on his behalf at the Royal Society. In 1788, 'On the Probabilities of Survivorship between Two Persons of Any Given Ages, and the Method of Determining the Values of Reversions Depending on Those Survivorships'.[18] And in 1789, 'On the Method of Determining, from the Real Probabilities of Life, the Value of a Contingent Reversion in Which Three Lives are Involved in the Survivorship'.[19] These two papers with their sober titles had a momentous result. In 1789 William was awarded the Copley Medal. The Copley Medal is the Royal Society's oldest award, given annually since 1731 for outstanding achievements in either physical or biological sciences, and still its most prestigious accolade. Later recipients include Charles Darwin, Albert Einstein, Jean Foucault, Michael Faraday, Francis Crick and Stephen Hawking – pioneers in the scientific world whose names have fame beyond academia and even a touch of glamour.

How was William's work pioneering? He certainly did not invent pensions or life assurance, both of which have classical roots.[20] Greek and Roman texts, also biblical references, show that the granting of a pension was an ancient practice – similarly burial societies, where funeral expenses were covered by members' subscriptions. And the concept of insurance against fire, loss of cargo on a sea voyage and other misfortunes can be found in the laws of early civilisations. In the Middle Ages there are examples of merchants insuring their lives, as well as their cargoes, but only for a limited period, usually a year. Life insurance as such began to emerge towards the end of the sixteenth century and, alongside it, the study of probability became a logical step. In the modern world Abraham de Moivre (1667–1754) is recognised as laying down the foundations of actuarial principles and, in particular, applying the theory of probability when determining annuities.

A form of long-term life assurance began as early as 1706 with a charter granted to the Amicable Society for Perpetual Assurance Office.

The Amicable rules precluded anyone over the age of forty-five from taking out life insurance so, when in 1756 James Dodson, aged forty-six, was refused a policy, he determined to develop a rational age-dependent scheme.[21] Using the London mortality tables Dodson identified not only the average number of deaths in any year but also the number of deaths in the worst year of the sample data. From these he created the *mean deaths table* and the *worst deaths table*. Using the worst deaths table Dodson built pragmatic caution into the premiums he proposed. He set out his conclusions in his *Lectures on Insurance*, alongside which he held a series of meetings at the Queen's Head in Paternoster Row – meetings which led to the formation of the Society for Equitable Assurances on Lives and Survivorship.[22] Dodson died in 1757 before the Equitable was formed, in 1762, but his *Lectures on Insurance* and the tables he constructed were crucial to the establishment of proper scientific principles for life assurance.[23]

William opens his first paper by disparaging de Moivre as outdated and 'incorrect', and erring 'most egregiously'.[24] Dodson, too, 'having derived his rule from a wrong hypothesis, has rendered it of no use'. His blunt dismissal of these two experts raises actuarial eyebrows even today. William does not spell out exactly where de Moivre and Dodson are at fault, but it is reasonable to conclude that he is referring to their assumption that the yearly rate of mortality is the same at all ages. In fact, mortality rates increase with age.

After dealing with de Moivre and Dodson in a few short paragraphs, William devotes the main part of each paper to formulae and tables. He demonstrates their use by means of problems:[25] 'supposing the ages of two persons, A and B, to be given; to determine the probabilities of survivorship between them from any table of observations'. And solutions: 'let *a* represent the number of persons living in the table at the age of A . . .'. It's specialist actuarial mathematics – and it works!

Writing in the *Times* more than a hundred years later, Sir William Elderton, an eminent twentieth-century actuary, described William's work as 'in its day, highly original'.[26] Elsewhere Elderton expanded his praise of William:[27]

He was the first to show how to work out complicated benefits, involving several lives, from any mortality table; the first to value the liabilities of a life assurance company and appreciate the meaning of the result; the first to see that, with the valuations in use, a margin of surplus had to be carried forward to prevent his bonus system from breaking down; the first to set down the available sources of profit and obtain measures for them; the first to keep record of the mortality of a life assurance office, and to notice that there was such a thing as 'select' mortality. Further than this, he was the first practical administrator of life assurance and a successful man of business.

Today William is known as the 'father of the actuarial profession'. What would he have made of this compliment? He regarded the title 'actuary' as an 'affected appellation'[28] yet he objected to people calling themselves actuaries when they were nothing more than 'schoolmasters and accountants'.[29] I suspect on balance he would have liked the paternal role and the respect which modern actuaries give him.

VERBAL FISTICUFFS

Nullius in verba – 'Take nobody's word for it'
(Motto of the Royal Society)

By 1790 William had presented four papers at the Royal Society; he had been awarded the Copley Medal, their highest honour, but he was not yet a Fellow. He very nearly scuppered his chances of election.

He had, in 1781, published a short work on heat and fire. It opens with a wide sweep:

> Fire, whether considered as an *element* or *quality*, so universally pervades all bodies, its properties and effects are so surprising, and its agency so subtile [*sic*] and powerful, that it is no wonder the researches of philosophers have often been directed to its investigation ... [but] notwithstanding the labours of this enlightened age, it still remains deeply involved in darkness and mystery.[1]

It is an eloquent expression of William's marvel at the world around him. And more. In the body of the paper there is passion, anger and even caustic wit.

The full title of William's publication is *An Examination of Dr Crawford's Theory of Heat and Combustion*, and it is just that – an examination (and a highly critical one) of a 1779 book: *Experiments and Observations on Animal Heat, and the Inflammation of Combustible Bodies; being an attempt to resolve these phenomena into a general law of nature.*[2] The author, Dr Adair Crawford, was a well-regarded physician and chemist, a friend of Joseph Priestley and respected in particular for the careful accuracy of his experiments.

Crawford's theory was that there is latent heat (he called it *absolute* heat) contained within the composition of all bodies, but this is not noticeable when they are in an atmosphere of the same temperature. Once, however, the latent heat of any object becomes greater than that of its surrounding medium it is obvious to sight and touch – in other words 'heat in the vulgar acceptance of the word'.[3] Crawford called this *sensible* heat.

William did not accept Crawford's theory and, in his paper, gives detailed descriptions of the Crawford experiments, which he repeated and expanded. His conclusion is damning: 'I think it must be acknowledged that his theory, so far as it is supported by his present experiments, is very defective and communicates but little real information.'[4] His final paragraph is dismissive:

> Upon the whole I wish Dr Crawford had employed his abilities in making a greater number of experiments, and a more thorough investigation of his theory before he proposed it. He should consider that ... much harm is done by such theories when wrong, by putting us back in the pursuit of truth, and occasioning a sad waste of that time and attention of philosophers in clearing away rubbish, which should be employed in raising and improving the superstructure of knowledge.[5]

His comments are a measure of William's intellectual curiosity and his search for scientific 'truths', but are nothing if not blunt. He finishes his paper with a nod to courtesy by declaring that he respects Crawford's character and accepts that he wants 'the advancement of science', and he adds a conciliatory note: 'should he, contrary to my expectations, by any new evidence remove my objections or establish his theory, I shall thank him, and retract any mistakes into which I may have fallen'.[6] Though veiled in polite language, this is a challenge – and it has a condescending tone which was not lost on Crawford. He was needled.

William Morgan had made an enemy.

Nine years later, in 1788, Crawford brought out a second edition modifying his *Theory*. The 'Welsh temper' which had put William's first employer in the gutter was roused. This was still a period when honour was defended with swords and pistols (even William Pitt fought a duel when he was Prime Minister) but the Morgan versus Crawford dispute took the form of verbal fisticuffs. Their spat, slogged out in the pages of the *Gentleman's Magazine*, was ostensibly masked by anonymity, but it was not difficult to deduce that the biting satirical tone of 'X. Y. Z.' came from the quill of William Morgan and that the righteously indignant 'A. B.' was Adair Crawford:[7]

Biff! A puny pamphlet has been enlarged to 500 pages but with so many alterations in his experiments and even in his theory that hardly a vestige of the original remains. Dr C. has spent nine years in a pursuit which has served no other purpose than to expose the errors of his first experiments. X. Y. Z.

Thwack! *Everything that can be considered as of the least importance continues* <u>unaltered</u>. A. B.

Wallop! When the results of any two experiments happen to differ from each other the discordancy is never regarded as an objection; but, with a happy facility peculiar to himself, he prefers the experiment which favours his hypothesis. X. Y. Z.

Sock! *The Critic has proved himself grossly ignorant of the first principles of the work he has undertaken to criticise.* A. B.

Bang! To enumerate all that is curious in Dr C.'s treatise would be to transcribe the whole book. X. Y. Z.

Jab! *The Critic has entirely perverted the meaning of the author.* A. B.

Smash! With a candour peculiar to himself he acknowledges that even some of his facts may be liable to error so it may perhaps be objected that the superstructure can hardly be secure when the foundations upon which it is supported are unsound. X. Y. Z.

Smack! *It would seem strange that enquiries, which are calculated only to illumine the understanding, should so frequently inflame the passions,* *and that envy and malice should follow the footsteps of those who are employed in tracing the harmony and beauty of the universe.* A. B.

Punch! We are almost led to believe that Dr C.'s system was originally discovered by the intuitive perception of the author and therefore wants not the precarious aid of experiments for its support. X. Y. Z.

Pow! *To a superficial observer it may appear a matter of wonder that personal attacks upon authors, acrimonious language, or malignant representations, should ever mix themselves with the discussion of questions in philosophy which are merely speculative.* A. B.

Thump! The gradual approximation which Dr C. is continually making to the truth is very amusing. X. Y. Z.

Swipe! *In the great mass of mankind certain spirits will always be found who look with a malignant aspect on the success of their contemporaries and who take delight in preying upon character.* A. B.

Crawford's article struck the final blow in the verbal brawl and there the matter appeared to have ended, but Crawford's anger continued to simmer. When in 1790 William Morgan was proposed as a Fellow of the Royal Society, Sir Joseph Banks, the President, received an anonymous letter – thwack again – objecting to the election of the 'nephew of a celebrated dissenting patriot who, presuming on the acquisition of a medal, is emboldened to become a petitioner for more constant honours'.[8] Banks showed the letter to Price who had also received an anonymous letter. Price's letter to Banks makes it plain that he is certain the letters come from Crawford or his friends: 'It is not conceivable that Mr Morgan can have any other enemy in the society, or, indeed, in the world.' Price writes with pride about William – 'an able experimental Philosopher; and one of the first Mathematicians to whom one particular brand of mathematics is greatly indebted'.[9]

In a second letter[10] Price is robust in his defence of William's remarks in the *Gentleman's Magazine*: '[t]hey contain no other invective against Dr C[rawfor]d than a representation of the inconsistency between the two editions of his pamphlet, the incorrectness of his Experiments, and the absurdity of his Theory'. Given the tone of

William's articles, Price's response is generous. If, as is likely, there were further anonymous letters, they did not deter the twelve Fellows who proposed William's admission. It is a star-studded list of distinguished men of science: Andrew Kippis, whose *Cook's Voyages* gave an account of James Cook's exploratory journeys to Australia, New Zealand and Hawaii; Abraham Rees, the compiler of a cyclopaedia which in forty-five volumes covered a wide range of topics in the fields of science, technology, agriculture, banking and transport, as well as the arts; Nevil Maskelyne, the Astronomer Royal, who developed a method of using the moon to calculate longitude, and established the *Nautical Almanac*. Another signatory was Francis Maseres, a lawyer and mathematician who had proposed a scheme to provide annuities for 'Britain's industrious poor'.[11] There was Sir James Edward Smith, botanist and founder of the Linnean Society;[12] Henry Cavendish, whose pioneering research covered a wide field of physics;[13] and Sir Charles Blagden, Secretary of the Royal Society and Copley Medallist of 1788. Their citation read: [14]

> William Morgan Esqr of Chatham-Square being desirous of the honour of admission into the Royal Society, we[15] whose names are underwritten, knowing his distinguished abilities as a Philosopher and Mathematician, of which he has given proof by three communications to this Society for the last of which he has received the prize medal of last year, do recommend him as highly deserving of that honour.

From its modest beginnings in 1660, when Christopher Wren and eleven others formed a society for 'the promoting of experimental philosophy',[16] the Society (Royal in 1662) had become a major presence in the scientific world. A society whose members collected 'the Common or Monstrous works of Nature' – plants, insects, seeds, bones, shells, feathers – originally hoping to gather 'into one room the greatest part of all the several kinds of things, that are scatter'd throughout the universe'.[17] A society whose members made detailed drawings, botanical, zoological and anatomical, and, increasingly, a society whose members conducted systematic experiments which they reported and published in the *Philosophical Transactions of the Royal Society*.

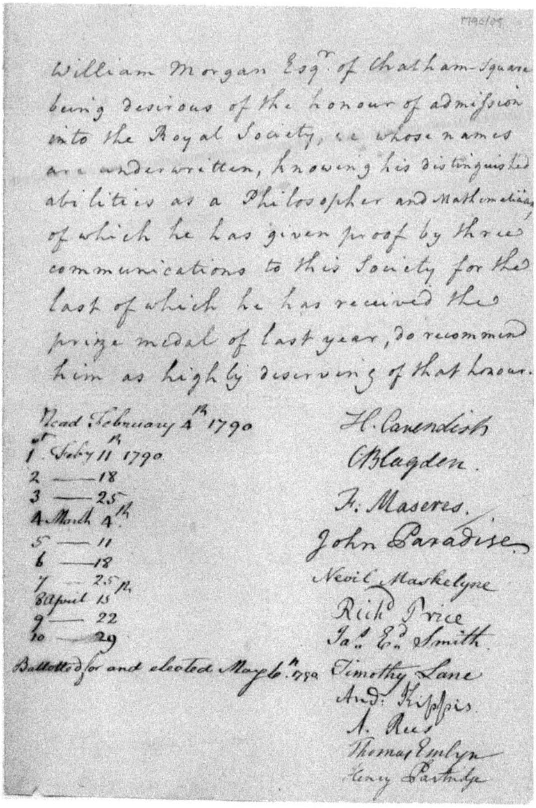

FIGURE 15 William Morgan's certificate of
election to the Royal Society, 6 May 1790
(*The Royal Society*)

This was the heady world, located since 1780 at Somerset House in the
Strand, which William joined on his election on 6 May 1790. True, there
were already criticisms that the Society was something of a gentlemen's club
with membership drawn predominantly from the landed aristocracy and
the clergy – those with sufficient leisure and money to pursue their interests. Within William's lifetime grumbles became more outspoken, notably
with the 1830 presidential election of the Duke of Somerset, someone
with no qualifications other than his being affable and royal. William,
always serious about his work, would have been pleased that eventually
scientific method and the pursuit of excellence trumped amateurism.

STAMFORD HILL

A splendid view indeed; at last, the richest
jewel in a monarch's crown is mine.
(James I, viewing London from Stamford Hill[1])

By the time of his Copley Medal award and his Fellowship William was married with a young family. There is no record of the date when his aunt, Sarah Price, introduced him to Susanna Woodhouse but they were married in 1781. The Woodhouses were from the Midlands and, by virtue of landed property, were a family of some standing. Susanna's grandfather, John Woodhouse, had died a wealthy man, leaving estates in Leicestershire, Shropshire, Staffordshire and Worcestershire. A Nonconformist minister and the founder of a Nonconformist academy, he was twice jailed for his views.

Susanna's father, another John Woodhouse, born in 1709 and living until 1771, was a 'merchant of London and of Portway, Staffordshire'.[2] It seems likely that he shared his father's religious views, given that his children were baptised in Nonconformist meeting houses. Susanna's mother, Hannah Woodhouse, by 1781 widowed for ten years, was 'an old friend and connection' of William's aunt, Sarah Price.[3]

What of Susanna herself? Born in 1753, the fifth of John and Hannah's six children, she had inherited the small estate of Portway in Staffordshire, a very modest part of her grandfather's landed empire. Caroline Williams gives no details about her looks and character other than that she was 'prudent and capable'. This smacks of faint praise and tells us very little – she sets a problem for a biographer.

William was, by the time of his marriage, thirty-one; he was serious, hard-working, and successful. In May 1781 his salary of £150 was doubled and the increase backdated. In addition he earned consultancy fees 'by answering annuity questions' and his name became widely known.[4] In 1786, for example, a dispute over lease-for-life tenancies on land held by the bishop of Exeter meant that William was called upon to give the valuations.[5] His judgement was respected; his prospects were good.

William and Susanna were married on Thursday 8 November at St Bartholomew the Great at Smithfield. Large and handsome, it is one of the oldest churches in the City of London and was once part of a priory which also housed St Bartholomew's Hospital. I visited on another November day more than two hundred years later. Outside, yellow sycamore leaves were skittering on the pavement; inside, an organist was playing a Bach toccata, the music leaping up to the massive Norman arches, whilst below, the air was scented by pedestals of cream and white flowers. Clearly these were preparations for a wedding – today it's a popular and fashionable place to be married and it was used for the final wedding in *Four Weddings and a Funeral*. It is surprising to picture William and Susanna in such a grand place, but the law required an Anglican service and St Bartholomew's was Susanna's parish church. Fortuitously, it was sympathetic to Dissenters; John Wesley, banned from other London churches, preached there in 1747 and 1748.[6] Another link with the recent past which would have pleased William was the fact that the Lady Chapel had once been used as a printing workshop, and Benjamin Franklin had worked there when he first came to London in 1725.

Their marriage was witnessed by Richard Price and Susanna's brother-in-law, Samuel Crompton.[7] The marriage certificate does not show who else attended the ceremony, in particular whether Nancy was there.[8] When William married she was no longer needed as his housekeeper and she returned to live with her mother and youngest sister, Sally, in Bridgend. Kitty, the eldest of the Morgan sisters, having in 1773 married the dashing Doctor Jenkin Williams, also lived in Bridgend.

Before long Nancy was married. Unlike the romance of her sister Kitty, with her misspelled billets-doux, Nancy's marriage is hard to depict as a love match. Her husband was Walter Coffin and his gloomy surname seems absurdly appropriate. Walter Coffin was a widower almost twenty years her senior, a quiet reserved man whose way of life was severely frugal. His father had wasted most of the family fortunes with crackpot speculations; his eldest brother, on inheriting, had squandered the remainder before his death. The next brother, also extravagant, died young, leaving Walter with the responsibility of caring for his mother in very reduced circumstances. He was a tanner and, by hard work and an extremely thrifty and self-denying lifestyle, had become a wealthy man.[9]

Nancy knew that, though marriage to Walter Coffin would make her rich, life would have to be quiet and simple – very different from the busy time and stimulating company she had enjoyed in London. At first she hesitated. She was twenty-eight; perhaps she felt there would not be any further chances to marry. Perhaps she was swayed by a letter from her uncle, Richard Price. 'May Heaven direct you to what will make you most happy', he wrote to her. 'But I hope, should you yield to Mr Coffin's wishes, that our loss [in her leaving London] will prove your gain by uniting you to a rational, good-tempered, worthy and generous man. No man deserves you who is not of this character.'[10] Or perhaps in the end she recognised for herself that this solemn man with his obsessively careful way of life was honest and trustworthy – and sincere. They were married at St Illtud's, Bridgend, on New Year's Day 1782.[11] This was only two months after William and Susanna's marriage, and suggests a whirlwind romance, but is more probably a reflection of Walter Coffin's long-held regard for Nancy.

George had by this time moved to Norwich, until the twelfth century second only to London in size and importance, and still a 'fine old city ... with its venerable houses, its numerous gardens, its thrice twelve churches, its mighty mound'.[12] As well as the 'thrice twelve' churches, there were a number of meeting houses for Dissenters, amongst them the very fine Octagon Chapel, attended by a wealthy and highly literate congregation. Here, George became the Unitarian Minister.

Norwich was the hub of a busy network of coaches, carriers' wagons and water transport. Wherries on the river Wensum reached outlying villages, while regular barges travelled to Great Yarmouth. One of the merchants of Yarmouth was William Hurry – 'wealthy, hospitable and generous'.[13] On 7 October 1783 George married Ann, the eldest of the Hurry daughters, a marriage which connected him with one of the most prominent Dissenting families in Norfolk.

In London Richard Price's wife, Sarah, was becoming increasingly frail and limited by her 'paralytic disorder'.[14] She died on 20 September 1786 at the age of fifty-eight. She had been an invalid since 1762, nearly half her life, and her death was not unexpected. This did not diminish Price's grief at his loss; he wrote to his sister of his 'inexpressible anguish at this separation after thirty years of uninterrupted happiness'.[15]

He could not bear to be in the house at Newington Green and he turned in his distress to his two nephews, first fleeing to be with William and Susanna, who were by now lodging in Sydenham. In a letter to George, written just a week after Sarah's death, Price told him, 'Your Brother has been my great support and refuge in this calamity; nor do I know how my weak spirits could have got thro' it without him.' In the same letter he welcomed an offer from George to visit him: 'Indeed I can scarcely think of anything that would have a greater tendency to comfort me than your company added to that of your Brother and his family.'[16] His mental pain was debilitating, and he was uncertain whether he would be able to resume his public duties.

The Marquis of Lansdowne, an old friend of Price's, offered him a period of quiet seclusion at Bowood, his country estate in Wiltshire, but he refused.[17] Nevertheless, his letters to the Marquis of Lansdowne show his spirits recovering and at the same time his ability to return to his studies and his preaching. By 9 October, barely three weeks after Sarah's death, he was able to plan for the future and he asked his sister to come to London to keep house for him.

The sister invited by Price was Sarah Morgan, mother to William and George. In March the following year Price moved to Hackney, little more than a village until early in the nineteenth century, and Sarah Morgan (together with her – as yet – unmarried daughter, Sally), was installed as housekeeper and companion. Aged sixty-two, Sarah Morgan had spent all her life in Bridgend, but the move to London was a surprising success. 'My mother seems to enjoy as much health as usual, and to be upon the whole happier than I expected', wrote William to his sister Nancy. 'They [Sarah and Sally] are out almost every afternoon or engaged with company at home.'[18]

The marriages of William and his siblings were followed by a new generation of Morgans. Kitty and Jenkin had seven children, Nancy and Walter had five. George and Ann had a daughter (Sarah), then eight sons, arriving at two-year intervals until 1799. William and Susanna had two daughters, Sarah and Susan, then four sons: William, John, Cadogan and Arthur. But the birth dates reveal how slowly William and Susanna's family grew. They were married for five years before Sarah arrived and the age gaps between their children vary from one year to six, with the births spreading over seventeen years. It seems very likely that there were stillbirths or infant deaths but I have been unable to find any baptismal or other records.

There is something else that is surprising and suggests that the bare facts of the family tree may hide heartache and loss. Arthur introduced a new name into the family lexicon; all the other children, like their cousins, had names which appear throughout the generations on the family tree, but there is one name which is conspicuously absent amongst William and Susanna's sons: Richard. Both Kitty and George had sons called Richard Price. Why didn't William remember and honour the uncle who had been such an important influence in his life? Or was there a son called Richard who died in infancy?

Whatever the difficulties, William certainly expected to have a large family for whom he would need a large house, and in 1787, successful and prosperous, he felt secure enough to commission such a house to be built. He took a ninety-nine-year lease on a parcel of land in Stamford Hill and his was one of the first houses to be built in what was then still a rural neighbourhood – though little more than six miles north of the busy city of London on the old Roman road to York. In addition to lawns and flowerbeds, there was enough land for a good kitchen garden, paddocks and stabling for horses, and room for some livestock – chickens and pigs are mentioned in family letters. From the house, there was an uninterrupted view of the surrounding country.

There is no view of the countryside today, although Stamford Hill Road is wide and lined with horse chestnuts and plane trees. William's house no longer stands but, among the supermarkets and modern blocks of flats, there are large and expensive houses dating from the late eighteenth and early nineteenth century; the Morgans soon had neighbours also escaping from their workplaces in the City. In the early 1800s Nathan Rothschild lived immediately next door, an early resident in what became (and remains) a predominantly Jewish area. He and William had common interests in the world of finance and both men were forthright when they disagreed – which is probably why they were reputedly very good friends.[19]

Stamford Hill House remained in Morgan family ownership until the 1870s. A photograph taken in the late 1850s shows a substantial building, the main part of which is on four floors.[20] High sash windows with louvred shutters (the shutters would prove to be a wise precaution) look out on a sweeping lawn bordered by shrubs and mature trees.

FIGURE 16 Photograph (late 1850s) of William Morgan's house at Stamford Hill, built for him in 1787 and remaining in the Morgan family until the 1870s.

It appears to be comfortable and commodious but, not surprisingly, it was expensive. William, ever careful in money matters, was so horrified by the builders' bills that he was almost driven into a fever.[21]

At the same time as William moved to Stamford Hill, George returned to London, initially moving to Hackney to be near the widowed Richard Price. Here he assisted, and hoped to succeed, his uncle as preacher at the Gravel Pit Chapel. The congregation, however, did not want him probably because 'his theological opinions were ... too pronounced to be generally acceptable'.[22] This is something of a euphemism – George's views, political as well as theological, were becoming ever more radical. William was indignant at the rejection of George, an indication that he shared his brother's views or saw nothing alarming in them.

At Southgate George ran a small school, taking pupils from 'liberal families', some of them following him from Norfolk. He also pursued scientific research with a fervent enthusiasm (he often rose at four in the morning to conduct experiments) which bursts from every page of the two volumes of his 1794 *Lectures on Electricity*. George and Ann's house at Southgate was five miles from Stamford Hill, close enough for the families to walk across the fields to visit each other and for the brothers to compare their scientific experiments and to discuss the political situation. On the national and international stage turbulent times were approaching.

10

A STUPENDOUS EVENT

Tremble all ye oppressors of the world! Take warning all ye
supporters of slavish governments and slavish hierarchies!
Call no more (absurdly and wickedly) reformation, innovation.
You cannot now hold the world in darkness. Struggle
no longer against increasing light and liberality. Restore
to mankind their rights and consent to the correction of
abuses, before they and you are destroyed together.

(Richard Price[1])

'The King's indisposition is a sad calamity and throws us into a state of great confusion and danger', wrote Richard Price in January 1789.[2] George III was mad. Today there is still debate about a diagnosis of his 'indisposition', though most medical historians think the likely cause was acute intermittent porphyria, a hereditary disease, exacerbated in George III's case by his diet, deliberately modest but as a result deficient in carbohydrate and glucose.[3] His eighteenth-century doctors, faced with an incoherent and deranged king, were at a loss and had to resort to a straitjacket to prevent him from injuring himself. His illness was a political as well as a personal catastrophe, for George III's incapacity made it necessary to appoint a regent, and the candidate for regency was the Prince of Wales – clever but headstrong, profligate and dissolute. By 1787 his debts amounted to £370,000, a cause of friction between father and son, and between parliament and prince.[4] Pitt had the unenviable task of negotiating terms and persuading the Commons to provide £221,000 to reduce the debt, although he knew that the prince, as soon as he attained power, was likely to dismiss Pitt and select another first minister from among his drinking cronies. Price was suspicious of Pitt, whom he judged as 'aspiring and ambitious', but he saw more danger in the 'loose and dissipated character' of the prince's supporters, notably Charles James Fox and Richard Brinsley Sheridan.[5] It was an unstable period for the country.

Pitt played for time, but by 5 February 1789 he could delay no longer and he introduced a Regency Bill into the House of Commons. On 16 February the Bill went to the House of Lords. Then, on 17 February, the king began to recover, and by 23 February he was lucid and coherent; there was no longer any need for the Regency Bill. On St George's Day the king and queen went in grand procession to St Paul's for a thanksgiving service and were cheered on their way by joyful crowds; George III's illness had inspired fondness for the monarch.[6]

The public's anxiety about the king's ill health and subsequent rejoicing about his restoration to sanity were the main news topic, but Price was observing events in France. His correspondence with Thomas Jefferson, who was American Commissioner and then Minister in France from 1784 to 1789, tracked the fermenting discontent across the Channel.[7] French support of the American colonies had won them victory over their old enemy, England, but at a very high price and their national debt threatened to bankrupt the country.[8] A feudal system and oppressive taxation already made for grinding poverty, which was exacerbated by two years of poor harvest. At the tip of the feudal triangle was Louis XVI, an amiable and cultivated young man but a weak and ineffective king. Below him was the first estate, the clergy, and below them the second estate, the nobility, both groups enjoying tax-free wealth and privilege. At the bottom of the social structure came the third estate, the middle classes and peasantry, frustrated by corruption in every part of the system.

A succession of finance ministers failed to rescue the precarious French economy and so in August 1788 Louis XVI called the Estates-General for the following year (the first gathering of representatives from all three estates since 1614), but the resulting assembly of 1,200 men lacked a structured agenda. The third estate seized their opportunity and declared themselves to be a National Assembly and, at a meeting in the tennis court at Versailles, swore they would not disband until France had a written constitution.[9]

In his *Memoirs of Richard Price* William claims that '[o]f all the events which distinguished Dr Price's life, none interested or agitated him so much as the French revolution',[10] so it is surprising that he makes no mention of his brother, George, being an eyewitness in Paris on 14 July 1789. The answer seems to lie in his bitter disappointment about the outcome of what he describes as 'this stupendous event – so auspicious in its beginning – so dreadful in its progress'.[11] He refers briefly to the Price/Jefferson correspondence and their 'judicious conjecture of the probable issue of the disputes between the Court, the Clergy, the Noblesse, and the Tiers État', but he does not include any quoted passages from them, 'subsequent events having long ago either fulfilled or falsified [them], and the subject itself affording but little consolation to the friends of peace and liberty'.[12]

Until very recently there was no full record of George's time in Paris, only snippets quoted by Caroline Williams from his letters home. The letters are lost. Then in 2008 Paul Frame, working on his biography of Richard Price, found them in the Newberry Library, Chicago, a discovery as stupendous as the events they describe – but tantalising.[13] Two remain missing. These are the ones he wrote to Price and which were published in the *Gazetteer* of 13 August and 14 September 1789. Mysteriously no copies of these issues have yet been traced, which suggests that they may have been suppressed.[14] The search goes on, but meanwhile the newly discovered letters recreate the scenes which William, disappointed and disillusioned, could not bring himself to include in the *Memoirs*.

George sailed to Calais on Friday 3 July 1789, together with three companions: a Norwich physician called Dr Edward Rigby, and two younger men, possibly ex-pupils of George's school, twenty-three-year-old Samuel Boddington and Olyett Woodhouse, who was twenty. The trip was in part Grand Tour but educational in a wider sense – a chance to observe different terrains, different agricultural methods, different cultures. And the expedition had a political colour; George carried letters of introduction from Price to some of those involved in the National

Assembly: Gui-Jean-Baptiste Target,[15] a lawyer who was one of the deputies of the Third Estate; and Mathon de la Cour, a mathematician and financier, to whom Price expressed his 'highest satisfaction' about the example of patriotism which France was giving the world: 'May Heaven grant success to this glorious struggle and make all nations free and happy'.[16] There was also a letter to Comte de Mirabeau, an aristocrat but an advocate of constitutional monarchy, whom William had probably met at the Equitable offices in the autumn of 1784.[17] George was, Price told Mirabeau, anxious for an introduction so that he could learn more about 'the glorious struggle in France for the blessings of Liberty'.[18]

Whatever 'struggle' the travellers might have expected they were told by the innkeeper at Calais that they 'should find Paris perfectly free from disturbance', and on their journey there were no signs of imminent uprising.[19] Rigby's letters paint a reassuring picture of life in France, 'What strange prejudices', he confesses,

> we are apt to take regarding foreigners! I will own I used to think that the French were a trifling, insignificant people, that they were meagre in their appearance, and lived in a state of wretchedness from being oppressed by their superiors. What we have seen contradicts this.[20]

George is more analytical. At Calais he observes that they 'never saw one person who appeared to be either drunk or dissolute', and in the villages they pass through, no evidence of 'indolence or dissipation'. But he notices 'scarcely any pasturage'; instead, crops of corn, beans and flax, and he reflects that France owes much of its strength to the hard work of the people, who feed 'on that subsistence which is given to the horses in Great Britain'.[21] When they reach the Prince of Condé's chateau at Chantilly George is outraged that 'an immense tract is wholly devoted to the insipid pleasures of one poor creature'.[22] In the lavishly contrived gardens he sees 'nothing but the mangled hands of Art . . . and a profusion of wealth without doing the least credit to the intellects of him who squanders it'.[23]

They reached Paris on 9 July 1789 and stayed at the Grand Hôtel du Palais Royal, each paying two guineas a night – in George's view 'a most extravagant lodging',[24] but one which proved to be both a safe refuge and a vantage point when the rioting began. The Palais Royal was a large square full of shops and cafes 'where crowds of the Parisians [met] every evening to talk politics, to drink coffee or to sup'.[25] From here they went sightseeing (Notre Dame, the Arsenal, the Bastille – George admitted to being bored) and on the weekend of 11 and 12 July they visited Versailles, where George delighted in the spectacle of 'the Grand Body of the Representatives assembled to establish Liberty in one of the first nations upon Earth'.[26] Rigby's letters record that they delivered the letters of introduction to 'Mirabeau, Target and some others of the popular characters in the Assembly', but that only Target was at home.[27] George's account of the meeting is likely to have been in one of the missing letters to Price. He makes no mention of it to his wife, but tells her that he was unimpressed with the 'splendid minions of a Court in all their gaudy tinsel'. He 'stared at the King and thought him a little superior to [their] own'. He 'stared at the Queen and saw the image of pride and lasciviousness'.[28]

Back in Paris they heard rumours of disquiet so they went to the theatre, hoping it would give them 'signs of the multitude's disposition'. Which it did. The manager announced that there was to be no play because the 'People' at his door 'had commanded it'. Outside crowds were gathering and 'every countenance expressed rage or the most anxious apprehensions'.[29] As they turned back to their hotel they could hear shouts of 'Aux armes!' and, with the night sky burning red with flames, the cries of the people mingled with gunfire. At ten o'clock they were shaken by a violent commotion as the mob stormed the gunsmith's shop just below them, where, by good chance, the raiders dispersed quietly when they discovered that the weapons had already been commandeered by another company.

In the morning their quarter of Paris seemed peaceful, but they were met with taunts, hisses, abuse and insults when they attempted to leave the city. After four days they secured a pass and travelled on, first to Dijon, and then via Lyons, Avignon, Marseilles, Nice and Turin to

Geneva. The journey was full of discomforts: bedbugs so bad George took to sleeping on the floor wrapped in a blanket; fleas so thick that they created a half-boot on Olyett's leg; and in Lyons 'a concentration of stinks. Garlick, rotten cabbages, fryes, roastings, inlets and outlets [from] a thousand sooty kitchens'.[30]

En route they were sometimes – waving their cockades – the first to deliver the news from Paris. In other places they were stopped by armed local groups but George was unalarmed, seeing in them the spirit of liberty rather than a threat to their safety. The letters end with a lyrical passage describing the beauty of Mont Blanc by moonlight and at sunrise, but the main thrust of all eight letters is in the political observations, typically George's comment as they leave Paris (after their four-day delay): 'the King's entrance without his guards into Paris, the demolition of the Bastille, and the restoration of Peace and Liberty to the noble Parisians amply repaid our loss of time and the fatigue of our spirits'.[31]

George's enthusiasm was echoed throughout Europe by the intelligentsia and by anyone who could read a news-sheet or join a cafe conversation. 'The papers contain such great and splendid news that I am hot from reading', wrote one German lady.[32] Further afield Kellgren, a young Swedish poet, wrote to his brother, 'Tell me, was there ever anything more sublime in History, even in Rome or Greece? I wept like a child, like a man, at the story of this great victory.'[33] And in St Petersburg Count Stroganov felt that he heard 'the cry of freedom . . . and the best day of my life will be that when I see Russia regenerated by such a Revolution'.[34]

In England, too, there was rejoicing in the news from Paris. Samuel Rogers wrote that the French Revolution was 'the greatest event in Europe since the irruption of the Goths'[35] and the mood was later famously encapsulated by William Wordsworth: 'Bliss was it in that dawn to be alive, / But to be young was very heaven.'[36]

Even the king and the governing classes could be pleased that the situation had, if nothing else, disabled a competitor and sometime enemy, though George III's pleasure was tempered by compassion for a fellow monarch. Pitt gave a cautious welcome to 'the present convulsions of France [which] must sooner or later terminate in general harmony and regular order' when 'France would stand forward as one of the most brilliant Powers in Europe'.[37]

For those campaigning for parliamentary reform in England, the disturbances in France were both a cause for celebration and a stimulus to renewed efforts at home. No one was more delighted than Richard Price. By now sixty-six years old, he was becoming frail, but a holiday at Southerndown at the beginning of September refreshed him. After a month of sea-bathing and rest, he returned to London feeling strong enough to accept an invitation to address the Revolution Society on 4 November 1789.

Revolution Society – the name might conjure up images of fierce desperadoes wielding guns, but the Society for Commemorating the Revolution in Great Britain, to give the Society its full name, was founded to remember a peaceful and political event rather than a violent and bloody one. Members met to celebrate the so-called Glorious (or Bloodless) Revolution of 1688, when the reigning King James II – a Roman Catholic – had been replaced by his Protestant niece, Mary Stuart, and her Dutch husband, William of Orange. For Price and the other members of the Revolution Society, Mary and William were monarchs chosen by the people rather than by God, and this was a reason for rejoicing.

William accompanied his uncle on 4 November when he gave his address – a discourse rather than a sermon[38] – at the Old Jewry Chapel in the City of London to a crowded congregation of old Dissenting friends and radicals including no fewer than four dukes (Norfolk, Richmond, Leeds and Manchester) and Pitt's brother-in-law, the Earl of Stanhope.

Price explored the idea of duty to one's country taking as his text Psalm 122, in which the Psalmist expresses 'in strong and beautiful language'[39] his love of country. For Price, country was not the 'spot of earth on which we happen to be born … but that community of which we are members'. Love of our country 'does not imply any conviction of the superior value to other countries' and he warned against rivalry between nations and a 'desire for conquest'.[40] True Christianity, he argued, should mean that 'animosity … among contending nations would be abolished.

Citing the parable of the Good Samaritan, he held that 'all men of all nations and religions were included in the precept, *Thou shalt love thy neighbour as thyself*.[41]

The country in which truth, virtue and liberty are found (and which the 1688 Revolution had helped to foster), Price preached, deserves the love of its citizens. On a personal note he expressed his delight that he had lived to see 'the rights of men better understood than ever', and he held up the Revolution in France in words which would later be seized upon by his opponents: 'I have lived to see thirty millions of people, indignant and resolute, spurning slavery, and demanding liberty with an irresistible voice, their king led in triumph, and an arbitrary monarch surrendering himself to his subjects'.[42]

He concluded in impassioned language, linking the French Revolution with the cause of further reform in England:

> I see the ardour for liberty catching and spreading . . . the domin-
> ion of kings changed for the dominion of laws, and the dominion
> of priests giving way to the dominion of reason and conscience . . .
> Behold, the light you have struck out, after setting America free,
> reflected to France and there kindled into a blaze that lays despot-
> ism in ashes and warms and illuminates Europe![43]

Price was, according to William, in considerable bodily pain when he delivered his sermon. Nevertheless, after a few hours' rest 'he exerted himself, though still weak and languid', and attended a dinner at the London Tavern, where he moved a Congratulatory Address to the National Assembly [of France].[44] At the dinner he was persuaded to publish his sermon. It appeared as *A Discourse on the Love of our Country* and became a bestseller 'read and admired with a fervour little inferior to that which had been heard in the Old Jewry'.[45]

But not by everyone. A review in the *Gentleman's Magazine* ascribed it to the 'dotage of Dr Price' and found the sentiment 'puerile' and the style 'vapid'.[46] William himself refers to a letter to Price from John Adams[47] in which he was contemptuous of the French Revolution and foretold its leading to the 'destruction of a million human beings'.[48]

Adams was incensed[49] that William had used the correspondence without his consent, his anger doubtless stoked by William's dismissive comment that his letter's 'harsh and gloomy censures . . . must have appeared as the effusions of a splenetic mind, rather than as the sober reflections of an unbiased understanding'.[50]

William's most furious outrage is directed at Edmund Burke, whose *Reflections on the Revolution in France* was published in December 1790. Burke attacked not only the *Discourse* but also a speech made by Price on 14 July 1790, when he acted as a steward at a dinner organised by Samuel Rogers on behalf of a new society, The Friends of the Revolution in France, and held at the Crown and Anchor tavern in the Strand. The Crown and Anchor was no ordinary tavern, in that it contained 'one splendid room measuring no less than 84 feet by 35 feet' and could accommodate as many as 2,000 people.[51] A throng of 652 attended the dinner organised by Rogers, a noisy celebration, not least when a waiter mounted the table and lifted up a block of stone from the ruins of the Bastille. Alcohol flowed as the company drank numerous toasts. Price made a short speech and invited his fellow diners to drink to 'an alliance between France and England to preserve universal peace and render the world happy'.[52]

Burke's pamphlet, three months later, was a bitter and personal attack on Price, on his interpretation of the achievements of the Glorious Revolution and on the welcome he gave to the French Revolution. Burke, William felt, behaved 'as if possessed by some daemon of the nether regions' in the 'torrents of abuse' and 'rancorous invectives' he had poured forth in his *Reflections*, and he deplored the damage which Burke had done to the cause of parliamentary reform: 'The phantoms which [Burke's] disordered imagination had raised to alarm and inflame the members of the House of Commons, unhappily succeeded too well in misleading the more timid and lukewarm friends of liberty'.[53] Moreover, Burke's antipathy to the French Revolution was such that 'in the paroxysms of his rage he denies even the principles on which the English revolution was established'.[54]

Price himself addressed – in very temperate language – Burke's objections in a preface and footnote to the fourth edition of the

Discourse. By contrast, William's repudiation of Burke's *Reflections* has a youthful energy although he was writing a quarter of a century after the events and at the age of sixty-five.

Burke's language may have been inflammatory but it was not long before his views became more generally accepted. A satirical cartoon of December 1790 by Gillray suggests that enthusiasm for the Revolution was waning. Called *Smelling Out a Rat; – or – the Atheistical-Revolutionist disturbed in his Midnight 'Calculations'*, it shows the huge nose of Edmund Burke 'smelling out' a threat to order and government in the person of Richard Price, who sits at his desk working on a pamphlet *On the Benifits [sic] of Anarchy Regicide Atheism.*[55] Above him, reinforcing the suggestion that Price favoured regicide, is a picture of the execution of Charles I, subtitled *the Glory of Great Britain*, whilst at his feet is a copy of the famous sermon which provoked Burke's attack. Gillray was in the pay of the government as early as 1788 (and intermittently thereafter), so that his cartoons were as often propaganda as they were a reflection

FIGURE 17 *Smelling Out a Rat; – or – The Atheistical-Revolutionist disturbed in his Midnight 'Calculations'*, by James Gillray (1790).
(Library of Congress)

of current opinion.[56] *Smelling Out a Rat* may have been commissioned, but Gillray injects some ambiguity into the satire: his explicit criticism is of Price and his progressive ideas, but the caricature ridicules Burke.[57]

Price did not live long enough to witness the way in which the French Revolution degenerated into mob rule and the guillotining of Louis XVI and Marie Antoinette. Looking back in 1815, William took comfort that his uncle was spared knowledge of 'those atrocious crimes which have stained the annals of Europe for the last twenty years'.[58] He claims, however, that Price would have felt justified by the declaration issued by the allied powers on entering Paris on 31 March 1814:

> The allied sovereigns receive favourably the wish of the French nation . . . they respect the integrity of ancient France as it existed under its legitimate kings . . . *they will recognise and guaranty the Constitution which France shall adopt. They therefore invite the Senate to name immediately a Provisional Government, which may provide for the wants of the Administration, and prepare the Constitution which shall suit the French people.*[59]

William's response to this declaration is a triumphant tribute to his uncle and a testimony to his own political views:

> No longer then let the servile and pensioned advocates of power traduce the memory of this virtuous man. No longer let them profanely dare to assert the right of kings to be divine, or derived from any other source than the free consent and suffrages of the people. Here we behold the greatest sovereigns of Europe compelled to acknowledge those rights which Burke had so peremptorily denied, and a Monarch whose ancestors had reigned over France for more than 800 years, raised to the throne, not on the ground of inheritance and indefeasible right, but under a new and solemn compact, founded on the unalienable rights of the people to choose their own form of government and to appoint those whom they should think fit to preside over it. Had Dr Price lived to witness these scenes, it would probably have given him some pleasure to

see his own principles established on the ruins of those of his great reviler; but it would have been a pleasure embittered by so much pain and mortification, that perhaps it has been happier for him to have quitted the world in the early period of the revolution.[60]

BUNHILL FIELDS

Whenever History shall rise above the prejudices which may for
a time darken her page . . . the name of Price will be mentioned
among those of Franklin, Washington, Fayette and Paine.
(*The Gentleman's Magazine*[1])

On Tuesday 19 April 1791, Richard Price died at the age of sixty-eight. He had become increasingly frail in the last year of his life. His regular trip to Bridgend in August and September 1790 boosted his spirits, but he did not have enough energy to work – as he had planned – on his memoirs. He spent the winter months in London but, feeling that his health was failing, he resigned his Hackney pulpit in February 1791. In the same month he attended a friend's funeral and caught a feverish chill which, despite a short-lived recovery, led to a 'complaint in the neck of the bladder'.[2] Price had bladder stones, which could be relieved only by 'surgical assistance'. William's discreet words refer to an excruciatingly painful procedure – an incision (without anaesthetic) between the scrotum and the anus, allowing the surgeon access to the bladder in order to remove the stones. This had to be repeated two or three times a day and then more and more frequently over the period of a month, until in the end it ceased to be effective.[3]

William charts Price's final hours with medical accuracy: his eventual acceptance that 'all was now over', his refusal of any further intervention by the surgeon's forceps, and the subsequent changes in his condition until his death, at a few minutes before three o'clock in the morning. The precise detail suggests that William was present but, in describing the deathbed scene, he slips suddenly into the third person, so it is not clear whether the 'nephew' he mentions is himself or George. The effect is to give a respectful formality to the occasion and a personal restraint. William does not write directly about his emotions, but his anguish can be inferred from his focus on the 'dreadful agonies' of the 'afflicted patient' and on his uncle's uncomplaining acceptance of his suffering.[4]

Price's death was an immeasurable loss to William and George. He had effectively been a parent to them both for the twenty years following the death of their father. He was their teacher, introducing them to the most up-to-date developments in mathematics and science – and to the leading mathematicians and scientists of the time. He was their moral mentor, involving them in theological and political debate with radical thinkers at home and abroad. In William's case, he launched him on a successful career. Above all, he was an example in how to lead a good life.

A week after Price's death his funeral procession left Hackney for the Dissenters' Graveyard at Bunhill Fields: 'six horsemen in their proper habiliments, immediately followed by nineteen mourning coaches some of which had four horses, besides a great number of gentlemen's carriages including the Duke of Portland, Earl Stanhope and other persons of distinction'.[5] The service took place at one o'clock.

This was not what Price had wanted. In the *Memoirs* William states that Price had made clear to both his nephews that his wish was for his funeral to be as private as possible and, accordingly, they planned a quiet evening service. In the event they were persuaded to have a day-time burial, which gave the opportunity for a large number to attend.

William was uncomfortable about Price's very public funeral service. His style gives him away. He continues in the third person, referring to George and himself as the 'two nephews' and 'the executors'. Here it is a transparent, though maybe unconscious, attempt to detach himself from his own conduct. In this section the first person is reserved for authorial disapproval: 'I am sorry to say', he writes, 'they suffered their better judgement to be overpowered by the solicitations of friends and admirers'.[6] His comment reads as if it is about someone else, not himself and George, and, given that there was an interval of twenty-four years between Price's death and the *Memoirs*, betrays an enduringly uneasy conscience. In his own and George's defence, he does allow that, by insisting that the funeral procession be kept from the most public streets, they prevented it from degenerating into 'a pageant very unsuitable to the remains of the modest and humble person who was to be the subject of it', but the tone is that of remorse. So who were the 'friends and admirers' who prevailed upon the two nephews and were the cause of so much self-reproach? They included the six Dissenting ministers who were his pall-bearers, amongst them Joseph Priestley, and, more clamorously, members of the Revolution Society, who published handbills giving the date and time of the funeral.

Andrew Kippis, the Dissenting minister and long-standing friend of Price, gave the graveside eulogy. The following Sunday, 1 May, Joseph Priestley delivered the funeral oration at the Gravel Pit Meeting House, describing Price as a 'benefactor of mankind' and focusing on the moral value of Price's work on the doctrine of annuities as well as his writings in the cause of civil and religious liberty. Priestley reminded the congregation that the National Assembly of France had 'justly styled [Price] the apostle of liberty'.[7]

The *Gentleman's Magazine*, not always sympathetic to Price, was warm in its tribute to him, praising 'his excellent understanding, his boldness and freedom of thinking, the purity of his views, and the simplicity of his manners' as well as 'that politeness and good-breeding which ever accompany native goodness and unassumed diffidence', and concluding:

> He was looked up to, and revered by, the Friends of Liberty throughout the world ... Whenever History shall rise above the prejudices which may for a time darken her page ... the name of Price will be mentioned among those of Franklin, Washington, Fayette and Paine.[8]

Bunhill Fields is an appropriate place for Price's grave. It is a green oasis just north of the City of London's square mile. A burial ground for over a thousand years (the name is thought to be derived from 'Bone Hill'), the site was preserved by an 1867 Act of Parliament as an

'Open Space' for the public to enjoy. Today the graves and memorials are protected by railings and padlocked gates; a network of paved paths intersects beneath plane trees, beech, horse chestnut, sycamore and lime. There are benches where office workers seek sandwich escape in their lunch hour, and here Paul Frame, Price's most recent biographer, first encountered Price's name.[9] I visited on a fine spring day; the daffodils were fading but there were grape hyacinths brightening the ground and the trees were greening. A huge plane tree arches over Price's table tomb, which lies just inside the City Road entrance. His uncle, Samuel Price, and his wife, Sarah, are also buried there, and Price rests in peace amongst a galaxy of other Dissenters, including Defoe, Bunyan, Blake, Godwin and Price's friend and fellow mathematician, Thomas Bayes. But the world he left behind was anything but peaceful.

In France, Louis XVI's failed attempt to escape to Austria and his arrest at Varennes on 21 June 1791 hardened hostility to the monarchy and, across the Channel, the news alarmed the English establishment, who viewed reformists with increasing suspicion. The political atmosphere was such that the Bastille Day anniversary dinner was a very muted affair compared with the triumphant celebration of the previous year. In Birmingham a similar dinner ignited the touchpaper for rioting and attacks on Priestley's meeting house and on his home. Priestley, warned not to attend the dinner, escaped and fled to London (and three years later emigrated to America), but the rioting continued with lives lost and buildings destroyed.[10]

And reputations stained. Reformists were labelled Jacobins. In France the name belonged to an organised and militant group whose leaders met near a Dominican monastery in the Rue St Jacques in Paris.[11] By 1792 they were in control of the revolution – the architects of the Reign of Terror and the merciless guillotine. The English Jacobins were different (and not to be confused with Jacobites) – intellectuals who believed in the power of human reason and to that end wanted freedom of expression. The name, they felt, was 'a stigma' fixed on them by their enemies.[12]

Reformists were even depicted – literally – as treasonous. A cartoon by Gillray titled *The Hopes of the Party, prior to July 14th – 'From*

such wicked CROWN & ANCHOR dreams, good Lord deliver us' shows
George III about to be beheaded, whilst Priestley gives him words of
comfort: 'Don't be alarmed at your situation, my dear Brother, we must
all dye once . . . a man ought to be glad of the opportunity of dying
if by that mean he can serve his country in bringing about a glorious
Revolution.' Another, *A Birmingham Toast, as given on the 14th of July
by the* ———— *Revolution Society*, shows Priestley, holding an empty
communion dish and an overflowing chalice, saying, 'The —— [King's]
Head, here!' Neither Priestley nor any of the men shown in the caricature was present at the dinner, but the damage was done.

Price, though lampooned in his lifetime, no longer appeared in
Gillray's cartoons but he was ridiculed in satirical verse.

> Let our vot'ries then follow the glorious advice,
> In the Gunpowder Legacy left to us by Price,
> Inflammable matter to place grain by grain
> And blow up the State with the torch of Tom Paine![13]

FIGURE 18 *The Hopes of the Party, prior to July 14th – 'From such wicked
CROWN & ANCHOR Dreams, good Lord, deliver us'*, by James Gillray.
(Library of Congress)

FIGURE 19 *A Birmingham Toast, as given on the 14th of July by the ———— Revolution Society.*
(*Library of Congress*)

In other doggerel swipes he was 'that Revolution-sinner Doctor Price'[14] and 'that Arch-devil, Doctor Price'.[15] The jibes veiled real danger; though friends and admirers leapt to his defence, Price was regarded as a dangerous hothead. Mary Wollstonecraft was one of the first to speak out, castigating Burke for his treatment of Price in her *Vindication of the Rights of Men*, but she published anonymously, perhaps in part a measure of the atmosphere of suspicion that prevailed.

William was outspoken in defending Price's good name (and at the same time attacking the government). In April 1792 he published *A Review of Dr Price's Writings on the Subject of the Finances of this Kingdom*, and was explicit about his two aims: firstly 'to do justice to the memory of an invaluable Friend',[16] and secondly to expose the government's failure to reduce the national debt, instead merely abolishing 'a few insignificant taxes' and 'imposing laws on posterity without the power of enforcing their execution, or even possessing sufficient virtue to make them the rule of our own conduct'.[17]

The main part of the publication was an account of three plans for redeeming the national debt which Price had provided for William Pitt in 1786. In his personal finances Pitt was unconcerned about living beyond his means. An annuity for his mother and a loan to his brother, together with his own lifestyle, meant that he amassed considerable debts. By the time of his resignation from office in 1801 these debts amounted to £46,000 (equivalent today to £2.75 million).[18] His attitude to the nation's money, however, was quite different and, when he came into office, he was determined to extinguish the national debt. The correspondence which William included in his pamphlet showed that Price had responded to a request from Pitt for advice by suggesting three possible courses of action. Each suggested that the government should put £1,000,000 a year into a Sinking Fund, and add to it with revenue from taxation. Price allowed for short-term borrowing in order to grow the fund by means of compound interest.

Pitt chose the proposal least favoured by Price, watered it down, and embedded it in the 1786 Sinking Fund Act. The scheme failed; far from reducing the national debt, it allowed government borrowing to grow – a situation exacerbated by the Napoleonic wars. Pitt cannot take all the blame for its failure. Price's biographers have acknowledged that there were flaws in his thinking, not least in his faith in compound interest. Carl B. Cone, for example, explains

> As a mathematical proposition, compound interest does make a small initial quantity increase with amazing rapidity. But government finance is quite a different thing. Only a clear surplus of revenue over expenditure can reduce a public debt. That surplus has to come from taxes paid either directly or indirectly by the people.[19]

D. O. Thomas and Paul Frame also acknowledge that Price bears some responsibility for the failure of the scheme though, as Frame argues, Price warned Pitt about the potential pitfalls. He was never shy about speaking out and, had he lived, he would undoubtedly have alerted the government to the emerging problems.[20] Ironically, as D. O. Thomas points out, the Sinking Fund Act gave Pitt credibility as a financial administrator and 'played a large part in the restoration of British financial credit'.[21]

Informed analysis of Price's plan was published as early as 1813 in Robert Hamilton's *An Inquiry Concerning the Rise and Progress, the Redemption and Present State and Management of the National Debt of Great Britain.*[22] It seems unlikely that William would have been unaware of Hamilton's criticism, but if it gave him pause for thought, there is no hint of it in any of his own publications. In all his later writing he remained as unswervingly loyal to his uncle as he was in his 1792 review of Price's plans. Over and over again it is the government at whom his anger and frustrated indignation are levelled. In 1792, he is particularly impatient with the government for obfuscation, and cites a Select Committee report in which 'a perplexed mass of estimates is thrown together, and the reader is led to imagine that, like the government of a kingdom, the public accounts are involved in so much mystery as to admit of being understood only by Ministers of State'.[23] He is critical, too, of what today we should call 'spin', in a Treasury report which, by selective reporting of debts and payments ('not the ordinary mode of balancing accounts'), took credit for a healthy economy, and he concludes by criticising 'vexatious taxes', such as the glove tax, hat tax and shop tax, the extension of excise on wine, tobacco and cotton manufacture, and three successive armaments.

> If ... the commerce of this kingdom has increased of late, it has been by the gradual operation of a peace of nine years, and the industrious spirit of the people, not by any encouragement it has received from the present Administration. Their claims to gratitude are indeed peculiarly improper, and they ought to blush in assuming to themselves the least merit on this occasion.[24]

12

A RADICAL FRIEND

Not wealth, nor power, can compensate for the
loss of that luxury which *he* has, who can speak
his mind, at all times and in all places.
(John Horne Tooke)[1]

Given his outspoken comments in his *Facts* and *Appeals* addressed
to the people of Great Britain, William was lucky to avoid caricature by the satirists of the day. Price and Priestley were prominent
targets and there were others who regularly appeared. Charles James
Fox, for instance, was a Whig who won his seat in parliament at the
age of twenty. He was a dissolute and extravagant womaniser, he was a
drinking crony of the Prince of Wales, briefly taking over 'Perdita', Mary
Robinson, the prince's first mistress.[2] John Wilkes was radical politician,
a wit and a rake – and publisher of the *North Briton*, the newspaper in
which, in issue 45, he denounced the king's speech as a tissue of lies.
Richard Brinsley Sheridan was playwright and poet as well as being a
Whig Member of Parliament and another drinking companion of the
prince. Thomas Paine was the author of *Rights of Man*. And there was
John Horne Tooke.

Fox, Sheridan and Horne Tooke all figure in Gillray's 1791 cartoons suggesting that the reformists wanted to overthrow (even kill)
George III. In *A Birmingham Toast* they are seated at the table, the worse
for drink, and ready to join Priestley's treasonous toast. In *The Hopes of
the Party* Fox raises his axe to strike the neck of George III, whose head
is held by Sheridan and legs by Horne Tooke.

If William's escape from caricature is surprising so too is his friendship with John Horne Tooke. There is a flamboyance about Horne Tooke's behaviour which sits oddly alongside William's sober and serious-minded outlook. But clearly both men valued their friendship very highly. William's preservation of mementoes from Horne Tooke is a measure of his regard for his friend. Evelyn Waugh in his autobiography mentions buttons embossed with the name 'Reform Club' which belonged to Horne Tooke and which had been handed down through the family.[3] In my case I have the lind stone and ribbon, the significance to William indicated by its careful safekeeping.[4] The meaning of this strange decoration was, however, to take some time to uncover.

John Horne Tooke, the son of a wealthy poulterer, was born plain John Horne, but he added Tooke to his name as a mark of respect to William Tooke, one of his political supporters (and perhaps in hope of a legacy from Tooke).[5] He was educated at Eton, where he lost the sight of his right eye in a brawl with a knife-wielding fellow student. (The injury does not seem to have damaged his looks; a later portrait shows a good-looking man with neat regular features and a steady gaze.) After Eton he went to Cambridge and, while still an undergraduate, enrolled as a member of the Inner Temple, the first step towards a career at the bar. He had a keen and logical mind and was an excellent public speaker. He would have been well suited to a legal career but his father bought him a living at New Brentford, worth about £200 to £300 a year, and insisted on his taking holy orders. He was not religious in any conventional sense and was cynical about church-going: 'In England people believe once a week – on a Sunday'.[6] Nevertheless, he did his best for his parishioners by studying medical books and opening a dispensary in his parsonage. He claimed that 'his medical labours were far more efficacious than his spiritual ones'.[7]

He never married. When an acquaintance suggested that he should reform and take a wife, Horne Tooke quipped, 'Whose wife?'[8] He had mistresses, however, and was father to three illegitimate children. Like many of his contemporaries he was quite open about this, and he made provision for his children. His two daughters made their home with him and he organised an annuity for their mother.

FIGURE 20 John Horne Tooke. Line engraving by
Anker Smith (1791), after a painting by Thomas Hardy.
(National Portrait Gallery)

He was lively and good company, his private conversation clever and
witty, overlaid with a veneer of nonchalant good-humour. A prominent
member of Samuel Rogers's social circle, many of his pithy remarks are
captured in the latter's *Table-Talk*. Hazlitt wrote of him that 'his intellect was like a bow of polished steel, from which he shot sharp-pointed
poisoned arrows at his friends in private, at his enemies in public', but
also that 'he said the most provoking things with a laughing gaiety and
a polite attention, that there was no withstanding'.[9]

William also had a fine mind and sharp wit, but the most obvious common ground between the two men is political. Horne Tooke was not especially concerned about the restrictions on Dissenters, but he championed liberty and free speech. An outspoken supporter of the colonists in the American War, he published, on behalf of the Constitutional Society, an advertisement for a fund for the widows and orphans of 'our beloved American fellow subjects', who had 'preferred death to slavery' and were as a result 'inhumanely murdered by the king's troops' (at the battle of Lexington). He was put on trial for criminal libel, fined £200 and imprisoned for a year.[10]

In 1780, after his release, he co-wrote with Richard Price a pamphlet called *Facts* and, with a grand sweep, addressed it to *The Landholders, Stockholders, Merchants, Farmers, Manufacturers, Tradesmen, Proprietors of every Description, and generally all The Subjects of Great Britain and Ireland*.[11] Published anonymously (though the authorship seems to have been an open secret), it contained an attack on North's government for its financial policies during the war with America. In a later (1796) publication of his own, William acknowledged his debt to the two authors of *Facts*: 'Dr Price and another learned and eminent patriot who is still living'.[12] Since Price was by this time dead, William's words imply that it was too dangerous to name the patriot who was 'still living'. In *The Memoirs*, published after Horne Tooke's death, it was safe to name – and praise – both Price and Horne Tooke. William noted how they had focused 'on the profligate manner in which the resources of the country were lavished, and on the ruin which it threatened to the liberties and property of the people', and had revealed corruption in the 'shameful bargains that were made with the different contractors for supplying the army and navy with stores and provisions'. Surprisingly, even Lord Shelburne, North's political opponent, objected to some passages and wanted the pamphlet suppressed. Tooke rushed to publication and in doing so started a quarrel with Lord Shelburne which 'admitted of no reconciliation during the remainder of their lives'. In *The Memoirs* William describes it as an 'admirable pamphlet', but this says little about the strength of their friendship.[13]

Like William (and Price), Horne Tooke was a member of the Revolution Society. At the 1790 anniversary dinner he joined the celebration of the French Revolution, though at the same time he showed his independence of mind. When Sheridan proposed a resolution giving a rapturous welcome to the establishment of liberty in France, Horne Tooke saw that this might be misrepresented as a call to revolution in England. In the face of booing and hissing he suggested an amendment recognising the existing and satisfactory constitution in England.[14]

Both Horne Tooke and William were members of the Society for Promoting Constitutional Information,[15] an innocent title and an organisation with peaceful reformist aspirations, but one which was to lead them into serious danger. The Society for Promoting Constitutional Information was founded in 1780 with the aim of achieving parliamentary reform by informing Englishmen 'what the Constitution *is*; what is its present danger; and by what means it may be placed in safety'.[16] As important as the Society's meetings were the books and pamphlets which it printed for free distribution, and which advocated measures such as annual parliaments, universal suffrage and voting by ballot. With election by ballot and the subscription at not less than a guinea, membership was exclusive: 'a number of members of Parliament and leaders of the reform movement, and a few very respectable gentlemen'.[17]

When William Pitt, son of the Earl of Chatham, Pitt the Elder, entered parliament in 1781 the reformists' hopes were high. Not only did his name carry weight but here was a leader with a gift for debate. Pitt moved a very cautious resolution calling for a select committee enquiry into the 'present State of Representation of the Commons',[18] but, whilst there was an appetite for economic reform to curb the expenses of the Crown, there was less enthusiasm for parliamentary reform. Pitt's motion was defeated by twenty votes, at the time regarded as a temporary setback, but in fact an outcome which proved to have been the nearest step to electoral reform until the First Great Reform Act of 1832, some fifty years later.

One immediate consequence of the rejection of Pitt's resolution was a Society for Promoting Constitutional Information meeting at the Thatched House Tavern in St James's Street. It was a venue they frequently used, but this meeting came to be known as *The* Thatched

House meeting – partly because it was decided to petition for a 'substantial Reform of the Commons' House of Parliament', but also, more significantly for subsequent events, because the decision was recorded in Pitt's handwriting.[19] It must have seemed harmless enough at the time but it was to embarrass him in years to come.

William's shared political sympathies with those of Horne Tooke do not explain the significance of his treasured memento of the ribbon and lind stone. There are no immediate clues in Horne Tooke's life story and none in his will, which is brief, leaving everything to one of his daughters and making no special bequests. The red ribbon is designed to be worn as a diagonal sash. Clearly it has a ceremonial function, whilst the lind stone with its pleated silk surround is effectively a rosette and has something of the cockade about it. Arguably and, given Horne Tooke's radicalism, it betokens some revolutionary organisation.

FIGURE 21 John Horne Tooke's regalia, possibly
worn at meetings of Jerusalem Sols.

Eventually the meaning of the sash and the stone is uncovered at Freemasons' Hall.[20] It is insignia, not Masonic, but probably belonging to a similar fraternity called the Royal Grand Modern Order of Jerusalem Sols, generally known as Jerusalem Sols, whose records show that Horne Tooke was one of their stewards.[21] Many such societies started at the end of the eighteenth century, with elaborate regalia and rituals, ceremonial processions and formal dinners. They all tended to have a legendary back-story, completely fictitious but creating the impression that the brotherhood was the inheritor of an ancient tradition. This, in an era when the king and the Church demanded deference, gave legitimacy to a secular organisation. In the case of Jerusalem Sols the society was dedicated to Solomon.[22] Horne Tooke's rosette with its goddess and cornucopia of flowers may have illustrated the Jerusalem Sols legend, perhaps depicting the riches of Solomon. By the end of the century, however, the government was so fearful of an English revolution and a French invasion that, in 1799, the Unlawful Societies Act was passed, banning all secret societies which required their members to take oaths. The Act, not repealed until 1967, was a death knell to Jerusalem Sols and numerous similar societies of the eighteenth century. Freemasonry's survival was thanks to its royal patronage, George III and six of his sons being Masons.

Jerusalem Sols had some 300 to 400 members in its heyday, but the remaining membership lists are incomplete and William's name does not appear in the existing records. Quite possibly he treasured the insignia as a memento of a friend rather than as a reminder of a society to which he belonged.

THE TRUMPET OF LIBERTY

The trumpet of Liberty sounds through the world,
And the Universe starts at the sound,
Her standard Philosophy's hand hath unfurled,
And the nations are thronging around.

(John Taylor: *The Trumpet of Liberty*[1])

By 1792 revolutionary fervour in France had transmuted into the savage bloodletting of the guillotine; in England the tide of opinion had turned against the Revolution and reformists were regarded with increasing suspicion. Wordsworth's response is typical. His much-quoted welcome to the Revolution, 'Bliss was it in that dawn to be alive', is in fact a retrospective memory disappointed by subsequent events. A hundred lines later he amends his earlier enthusiasm:

But now, become oppressors in their turn,
Frenchmen had changed a war of self-defence
For one of conquest, losing sight of all
Which they had struggled for[.][2]

Despite the climate of mistrust, the Society for Promoting Constitutional Information continued to applaud the principles of the revolutionaries and, in May 1792, sent a message of support to the Jacobin Society in France deploring the 'despotic ignorance' which had 'insulted their peaceable principles'.[3] With hindsight their reading of the situation seems remarkably naive, and unsurprisingly it contributed to the view that the reformists approved of violence.

Nervousness about the Society increased with their recommendation of Thomas Paine's 'masterly book',[4] *Rights of Man*, Part I of which was published in 1791, Part II in February 1792. Paine, already conspicuous for his pamphlets in support of the colonists in the war with America, was uncompromising in his views. 'All hereditary government is in its nature tyranny', he declared.[5] All men, he argued, are born free and with natural rights, some of which are adjusted when individuals join together to create societies. Society precedes government, and those who make up a society retain the right to change the government.

Here was a direct attack on the king and the House of Lords and, in the government's view, the book contained 'divers false, wicked, scandalous, malicious and seditious matters'.[6] Among the propertied classes in the country at large anti-Paine feelings ran high and in places effigies of him were burned. But the book, written in a colloquial and unpolished style, sold well, and the poor (those who were literate) must have relished his vision of a welfare state where taxation was fair and used to fund child allowances, education and old age pensions.

In the late summer of the same year William's brother, George, published *An Address to the Jacobine Societies*, opening with a rallying cry to his readers: 'Convince the Human Race, that to banish Kings and Courts, is to extirpate the most destroying Pestilence that ever desolated the Universe'.[7] In ten short chapters he delivered a blistering attack on Louis XVI, on hereditary monarchies and their governments. Unequivocally republican, he listed, in high-flown language, the evils of kings and aristocracy and was scornful of the so-called 'balanced power' of the English system. Much of the content was an echo of Thomas Paine's *Rights of Man*, and George might also have been charged with sedition had he not taken the precaution of publishing his work anonymously.[8]

William may seem to have been less fervent. He seldom attended meetings of the Society for Promoting Constitutional Information, perhaps because of pressure at work, perhaps because of domestic concerns – his first son, William, was born in 1791 – but he was not keeping a low profile. His *Review of Dr Price's Writings on the Subject of the Finances of this Kingdom*, with its criticism of the government, was published in April 1791, by which time he was already hosting Sunday evening gatherings. Given the prevailing climate of mistrust, his hospitality was defiantly brave; the parties, though social, had a political – and radical – edge. As a safety precaution the shutters were closed (William was well aware that the government was employing spies), but anyone snooping round the house would have heard the music and maybe recognised the words of *The Trumpet of Liberty*, a stirring song by John Taylor, a deacon at the Octagon Chapel in Norwich where George had been minister.[9] Everyone present joined in the chorus:

> Fall, Tyrants, fall, fall, fall!
> These are the days of liberty!
> Fall, Tyrants, fall!

The singing was led by Amelia Alderson, later Amelia Opie when she married the painter John Opie. Introduced by George, she was a friend from his Norfolk days; she wrote poetry and novels and was a vivacious and talented member of the Dissenting set.

Were Thomas Paine and William's long-standing friend, Samuel Rogers, guests? Probably, for both men enjoyed William's hospitality on at least one other occasion. In his diary for Friday 20 April 1792 Rogers recalls dining at William Morgan's, whom he describes in passing as 'a silent man, but very strong and emphatic in his language'.[10] The only other person mentioned is Paine. Paine's outspoken defence of Richard Price in *Rights of Man* would have been a prime reason for his friendship with William. In addition, the two men shared an interest in science – Paine was absorbed in matters as wide ranging as a design for a smokeless candle and one for a single arched bridge – but their chief common concern, and their focus at the dinner table, was the need for political reform. Paine by this time was a red-hot dinner guest.

William proposed a toast to the memory of Joshua .[11] Those present would not have needed reminding that the Old Testament book of Joshua records regicide on a grand scale: in one afternoon Joshua put to death the kings of Jerusalem, Hebron, Jarmuth, Lachish and Eglon, as a group, and the next day a string of others, one by one, with their followers.

It was by every standard a daring choice. Not only was William voicing a rejection of monarchy, but he was invoking the Bible to do so. Paine was more circumspect. 'I would not treat kings like Joshua', he declared. He preferred a prayer against Louis XVI which he attributed to a Scotch parson, 'Lord, shake him over the mouth of hell, but don't let him drop!' Paine in his turn gave a toast to 'The Republic of the World', in Rogers's view 'a sublime idea'.

A month later on 21 May 1792 Paine was charged with seditious libel. On 15 September he fled to France, where he had already been accorded citizenship, and he was tried *in absentia*. His defence counsel was Thomas Erskine, the lawyer who had successfully defended Lord George Gordon after the 1780 riots and who, in the intervening years, had become one of the leading barristers of the time. In arguing for Paine, Erskine delivered an impassioned speech for the freedom of the press, asserting that Paine was entitled to criticise the constitution as long as he advocated change by legal means. He also slipped in a reference to his own situation in having to defend an unpopular cause: 'Oh, gentlemen, where would the constitution be, if advocates were only to consult their inclinations, their foibles, or their virtues? It would be gone. Every case would be pre-judged, and there would be no occasion for the assistance of a jury in any case.'[12] His speech is still remembered by the bar as defining the 'cab-rank rule' – counsel cannot refuse a case simply because it is disagreeable. Neither his plea for himself nor his plea for Paine, however, persuaded the jury. They returned a guilty verdict without retiring.

Meanwhile Paine was given a 'flattering reception' throughout France.[13] He reached Paris on 19 September 1792, three days before the National Convention carried unanimously the motion that 'royalty be abolished in France'. A republic was declared and with it, in a symbolic start to the regime, a new calendar was introduced; 22 September became the first day of the month of Vendémiaire in L'an Un ('Year One') .[14] With the motto 'war on the castles, peace to the cottages', the Republic set about assisting all peoples to establish 'free and popular' governments. In other words, France, in the name of peace, liberty, fraternity and equality, was at war with most of Europe.

Then, in January 1793 (Pluviôse in the new calendar), Louis XVI was sent to the guillotine and beheaded. It is a historical fact so well known that it has lost the power to shock, but at the time the execution of the French king was an act which excited international horror. Pitt called it 'the foulest and most atrocious deed';[15] Tom Paine, despite his republican convictions, was appalled at what had been done; and for every Briton it was a chilling reminder of another January day, in the previous century, when their king – Charles I – had been beheaded.[16]

As well as precipitating further conflict (France declared war on England in February 1793), the execution of Louis XVI damaged, by association, movements for parliamentary reform in England. Reformists were seen as republicans (which many of them were) and therefore as intent on killing the king (which they undoubtedly were not). Radical societies were regarded with increasing suspicion. One which was to become prominent was the London Corresponding Society, founded in January 1792 by Thomas Hardy, a master shoemaker with a shop in Piccadilly. By charging only a penny a week subscription, Hardy hoped to enrol 'another class of people'.[17] The aims of the society were parliamentary reform and universal suffrage, members having to promise they would 'endeavour, by all the means in their power, to promote the objects the Society had in view'.[18] Other societies in London and the provinces flourished, but cooperation between the Society for Promoting Constitutional Information and the London Correspondence Society gave the government particular concern, especially when they set up a joint committee and began to plan for a convention. The word, however peaceably intended, had explicit echoes of the post-revolutionary regime in France and the government was on its guard. By the end of 1792 the Tower of London had been fortified and, in ten counties, militia mobilised.[19] Add rumour, spies and fear of invasion by France into the mix, and the atmosphere was increasingly jittery.

In 1793 several reformists in Scotland were arrested and, on flimsy evidence, tried for sedition. Found guilty, they were sentenced to transportation for periods varying from seven to fourteen years.[20] The fate of the Scottish martyrs, as they became known, far from deterring the English reformists, strengthened their resolve. Meanwhile in the country at large war was damaging commerce, which in turn caused unemployment and reduced wages. An anonymous pamphlet, *The Rights of Swine, An Address to the Poor*, published in 1794, drew attention to the condition of 'thousands of honest industrious people in Great Britain [who were] starving for want of bread ... whilst impudent nobles [were] advertising their Grand Dinners'.[21]

The mood was ugly and the government was alarmed. At six o'clock on the morning of 12 May 1794 Thomas Hardy, founder of the London Correspondence Society, was arrested and his house searched. On the same day the government intercepted a letter from Jeremiah Joyce (a Unitarian minister and a known reformist) to Horne Tooke. 'Dear Citizen. This morning, at 6 o'clock, Citizen Hardy was taken away by order from the Secretary of State's office. They seized everything they could lay hands on. Query, is it possible to get ready by Thursday? Yours J. Joyce.'[22] The message and, within it, the word 'citizen' with its revolutionary overtones seemed to confirm their worst fears. Horne Tooke later claimed that he was merely 'getting ready' some information about sinecures and pensions held by Pitt's cousins, the Grenvilles, but the damage was done.[23] On 15 May he was arrested and taken to the Tower of London. The following day the government began to rush through parliament an act suspending habeas corpus, in other words imprisonment without trial. The act received royal assent on 23 May, and within a couple of days there were twelve reformists being held in the Tower. They became known as the twelve apostles.[24]

To some extent Horne Tooke had himself to blame for his imprisonment. He knew that his membership of radical societies gave grounds for suspecting him, so, when a government spy became a visitor at his house in Wimbledon, he saw through the man's friendly overtures, and entertained himself by making false and preposterous confessions about conspiracies. He took the joke too far when he told the spy, under a promise of absolute secrecy, that an armed force was being organised and the country was on the verge of revolution.[25]

One can only imagine William's feelings when he heard of Horne Tooke's imprisonment. As well as concern for his friend, he had every reason to fear for his own situation when he learnt that 'his name was on a list of those threatened with prosecution'.[26] Then, in early July, William received the little slip of paper which I found carefully preserved in the box with Horne Tooke's Jerusalem Sols regalia – a subpoena to appear at Horne Tooke's trial.

GEORGE the third by the Grace of God of Great Britain France[27] and Ireland King Defender of the Faith etc To John Debrett John Stockdale William Morgan and George Morgan and to every of them Greeting We command you and every of you that laying aside all excuses and pretences whatsoever you and every of you personally be and appear before our Justices assigned to deliver our Gaol of Newgate of the prisoners therein being for our County of Middlesex on Monday the Twenty seventh day of October next by nine of the Clock in the Forenoon of the same day at Justice hall in the Old Bailey in the Suburbs of our City of London there to testify the Truth between us and Thomas Hardy Jeremiah Joice[28] John Horne Tooke Stewart Kydd and John Richter for high Treason on behalf of the defendants and this you or any of you are not to omit under the penalty of one hundred pounds to be levied on the Goods and Chattles Lands and Tenements of such of you as shall fail herein Witness Lloyd Lord Kenyon at Westminster the ninth day of July in the thirty fourth year of our Reign
By the Court Templer

FIGURE 22 Subpoena of 9 July 1794 summoning William Morgan, together with his brother George, John Debrett and John Stockdale, to the trial for high treason of John Horne Tooke and his co-defendants.

John Debrett[29] and John Stockdale were both publishers and book-sellers, risky businesses in the febrile atmosphere of the day. For all of them the subpoena spelt out the seriousness of the defendants' position. They were charged with high treason. If found guilty, the prisoners were likely to be hanged, or even hanged, drawn and quartered. By this time at least forty more people had been rounded up and charged with high treason, though not on the same indictment as the twelve apos-tles.[30] William was undoubtedly under surveillance (as well, probably, as his brother, George), but he continued to host his Sunday evening gatherings – a measure of his courage as well as his commitment to the reformist cause.

Horne Tooke had been arrested for 'treasonable practices' but, well versed in the law, he knew that there was insufficient evidence to mount a case against him. As late as 21 July he complained in his diary that he had been in close custody in the Tower for nine weeks 'without a possibility of forming the most distant conjecture concerning any charge to be brought against [him]'.[31] Given that he had no privacy, this naivety was deliberate and provocative. He had every reason to be angry about his situation, having spent the long hot summer of 1794 in the Tower. Iron bars were put up at his window and for the first week he was refused writing materials as well as newspapers. When he was allowed pen and ink and able to keep a diary he recorded details of his situation. He was provided with basics: tea and sugar, and lozenges for a cough, together with shirts, stocks, stockings and handkerchiefs, and he was given a weekly allowance of 13s. 4d (the equivalent today of some £60) from the government, but this was a reduction in the usual allowance for alleged traitors and insufficient to cover the cost of laundry and other expenses.[32] He had a close-stool – in other words, a chamber pot in a discreet wooden box – but, since there were always two warders in his room night and day, he had no privacy in which to cope with a long-term medical problem which he describes as a 'dropsy, or a rupture, or both, or a Schirrus[33] in the testicles, besides a more painful complaint in their neighbourhood'.[34] Horne Tooke's biographers have speculated about the condition causing these symptoms – possibly heart failure or renal disease.[35] Whatever the cause, Horne Tooke suffered considerable discomfort and he applied to the Governor asking that he might be attended by his physician, Dr George Vaughan, and his surgeon, Mr Henry Cline. After a month of discomfort his request was partly granted, and he received regular visits from a Dr Pearson and from Henry Cline. Physical relief – and, for Horne Tooke, the visits had an added bonus, in that Cline shared his political views.

Cline had been a contemporary of William's at medical school and they remained friends, or perhaps picked up again after an interval of years.[36] In a letter to his mother in 1803, William tells her that he spent 'an agreeable day at Mr Cline's'.[37] Given their political sympathies and their mutual concern for Horne Tooke, it seems likely that they would have been in touch in 1794 and William would have received regular news about Horne Tooke. Cline may well have been one of William's Sunday evening guests.

Outside the prison rumour spread. The word was that the government was so confident of conviction that they had prepared 800 warrants, 300 of which were already signed, ready for the immediate arrest of further suspects.[38] Speculation was fevered, and bets were laid on the lives of the prisoners. Brooks's betting book contains the entry 'General Fitzpatrick bets Charles Grey five guineas to one that Tooke is hanged before Pitt'.[39]

TRIAL FOR HIGH TREASON

My lord, it is not for a small stake that I stand here – it
is to deprive me of my life, to beggar my family, to make
my name and memory infamous to all posterity.

(John Horne Tooke[1])

The trial took place at the Old Bailey, which, rebuilt following the
destruction of the former building in the Gordon Riots and the
last word in modernity, was packed. Spectators, respectable working
folk alongside well-to-do intelligentsia, sat in high galleries with an
excellent view of the courtroom. The trial was at the same time serious,
theatrical and something of a social occasion, Samuel Rogers recording
that he paid five guineas for a loge[2] from which to view the proceedings and doubtless to nod to others who were attending.[3] Amongst
them was William Godwin, remembered today for marrying Mary
Wollstonecraft as much as for his radical political views. Godwin was
assumed to be (and eventually admitted that he was) the author of a
pamphlet published in the *Morning Chronicle* attacking the Lord Chief
Justice's address to the jury, in which, it was claimed, treason was being
inferred from other activities. If this was so, it was a doctrine that was
on trial as much as the prisoners. A more careful examination of the
judge's speech should have persuaded the writer that this was not the
case, but the pamphlet was widely read – and influential.

Amelia Alderson attended every day, reporting in a letter to
Mrs Taylor, her friend in Norwich, that George Morgan had read the
list of jurors with alarm; almost every man, he felt, was 'a rascal, and a

contractor, and in the pay of the government some way or another' and certainly none could be trusted.[4] George and various others were so gloomy that they were already turning 'a longing eye towards America as a place of refuge', and Amelia, too, was resolved 'to emigrate if the event of the trial be fatal'.[5] She was, however, she told Mrs Taylor, going 'as usual' to William Morgan's that evening and would sing John Taylor's song. The letter is an undated fragment, but the details about the jury and the fears for the outcome mean that it was written sometime in October.[6] In the circumstances William's hospitality was risky and brave.

The trial began with pomp and ceremony at one o'clock on Saturday 25 October 1794, when the Lord Mayor took his seat on the bench at the Sessions House of the Old Bailey. He was accompanied by six aldermen of the City of London, the Lord Chief Justice and four other judges. Counsel for the prosecution was led by the Attorney General, assisted by the Solicitor General and six other barristers. After points of order and adjournments the court reconvened on Tuesday 28 October and proceedings began.[7] The official record of state trials provides a verbatim report of the proceedings, transcribed from the shorthand notes of Joseph Gurney, and has generally been considered accurate.[8] But there is another account of the trial, published by Manoah Sibly, who was commissioned by the London Corresponding Society to make a shorthand record of proceedings. The accuracy of Sibly's reporting was challenged by contemporary reviewers, although he was credited with capturing the essence if not the exact wording of what took place.[9] A written account can never reproduce the nuanced inflexion of the spoken word but, taken together, the Gurney and Sibly publications provide an almost filmic picture of the trial.

The prisoners had elected to be tried individually and the first to be called was Thomas Hardy, the secretary of the London Corresponding Society. For four days counsel for the prosecution presented the case for the Crown: an opening speech of nine hours followed by documentary evidence and then the examination of witnesses. Minutes of meetings, reports of speeches and letters were produced as evidence that Hardy (and the other prisoners) had 'compassed the king's death'.[10] Only a spoof playbill, at worst in questionable taste and quite probably a plant by a government spy, could be said to have suggested killing George III.[11] The jokes and puns, even at a distance of over two hundred years, are obvious:

For
The Benefit of JOHN BULL.

At the
FEDERATION THEATRE, in EQUALITY SQUARE,
On Thursday, the 1st April, 4971,
Will be performed,
A new and entertaining Farce, called
LA GUILLOTINE;
OR,
GEORGE's HEAD IN THE BASKET!
Dramatis Personæ,
Numpy the Third, by Mr. GWELP,
(Being the last time of his appearing in that character)
Prince of Leeks, by Mr. GWELP, junior.
Duke of Dice, by Mr. FREDDY,
(from Osnaburgh.)
Duke of Jordan, by Mr. WILLIAM HENRY FLOGGER
(from the Creolian Theatre.)

Uncle Toby, by Mr. RICHMOND.
Grand Inquisitor, Mr. PENSIONER REEVES.
Don Quixote, Knight of the Dagger,
By Mr. EDMUND CALUMNY.
And Chancellor of the Exchequer, by
Mr. BILLY TAXLIGHT.
Municipal Officers, National Guards, &c.
By Citizens XOF, NADIREHS, YERG, ENIKSRE, &c.
Banditti, Assassins, Cut Throats, and Wholesale
Dealers in Blood, by THE EMPRESS OF
RUFFIANS, the EMPEROR OF HARM-ANY,
THING OF PRUSSIA, PRINCE OF S. CASH-
HELL, &c.
Between the Acts,
A new Song, called "Twenty more, kill them!"
By BOBADIL BRUNSWICK.

Tight Rope Dancing, from the Lamp-post,
By Messrs. CANTERBURY, YORK, DURHAM, &c.
In the course of the Evening will be sung,
in Full Chorus,
CA IRA,
AND
BOB SHAVE GREAT GEORGE OUR ———!
The whole to conclude with
A GRAND DECAPITATION
OF
PLACEMEN, PENSIONERS AND GERMAN LEECHES.
Admittance, Three-pence each Person.
Vive la Liberté! Vive la Republique![12]

Not until the afternoon of Saturday 1 November did Thomas Erskine, counsel for the defence, rise to make his address to the jury. After seven hours his voice was hoarse and he needed to lean on the table for support, but the courtroom was so still that even his faintest words were heard, and when he sat down there was a burst of spontaneous applause which was taken up by those outside the court. So many people thronged the streets that the judges were unable to get to their carriages. Eventually Erskine (his voice conveniently restored) had to go out and persuade the crowds to disperse.[13]

The examination of witnesses began later that evening and continued until after midnight, when the court adjourned. There had never been a trial like it and the outcome was uncertain. If Thomas Hardy were found guilty and hanged, what would follow for Horne Tooke and the other prisoners? What might happen to those such as William Morgan, whose names were on the threatened-with-prosecution list?

On 5 November, after the court had been sitting for nine days, usually from eight o'clock in the morning till midnight, the jury retired. Earlier in the proceedings Erskine had told them that, to find Hardy guilty of criminal intention, they must be convinced beyond all reasonable doubt – possibly the first recorded use of that now-familiar phrase.[14] After three hours the jury returned with their verdict: Not Guilty. The applause in the Old Bailey was so loud that it reached the crowds, waiting outside the court despite steady rain, and they greeted the news with 'the loudest acclamations of joy'.[15] Hardy, recalling the day in his *Memoirs*, wrote that 'like an electric shock, or the rapidity of lightning, the glad tidings spread through the whole town'.[16]

The streets were crowded again as Horne Tooke travelled each day from Newgate to the Old Bailey, a journey to which, years later, Samuel Rogers referred in his poem, *Human Life*:

> On thro' that gate misnamed, thro' which before
> Went Sidney, Russell, Raleigh, Cranmer, More,
> On into twilight within walls of stone,
> Then to the place of trial; and alone,
> Alone before his judges in array
> Stands for his life.[17]

The crowds wished him well, but Horne Tooke was well aware that, though Hardy's acquittal gave him grounds for hope, it was also likely to make counsel for the Crown even more determined to secure a guilty verdict in his case. His trial began on 17 November and, as soon as he was in the dock, he made an impassioned address to the judge. 'My lord, it is not for a small stake that I stand here – it is to deprive me of my life, to beggar my family, to make my name and memory infamous to all posterity.'[18] Next he used his poor health and his 'infirmity' to ask (successfully) that he be allowed to leave the dock and sit beside his counsel. Here, as reported in the *Gentleman's Magazine*, he 'assisted his counsel by pleading his own case with much animation'.[19] Delicate wording – Horne Tooke's frequent interruptions to proceedings were robust and skilful, and often witty.

Neither William nor George was called to give evidence. The very many witnesses whom Horne Tooke had subpoenaed included public figures whose appearance would have had much more impact than that of the Morgan brothers. There were MPs – Charles James Fox, Sheridan, Lord Stanhope – and even two cabinet ministers – the Duke of Richmond (Master-General of Ordnance) and the Prime Minister, William Pitt. For Pitt the examination was embarrassing. First, he had to admit to recognising his own handwriting when Horne Tooke produced a letter 'on the subject of parliamentary reform' – a sufficient reminder to all present that Pitt, now heading a repressive government, was once a champion of reform.[20]

Pitt had further to admit that he had been at the Thatched House meeting in 1782.[21] 'Were we not a convention? What was that meeting but a convention of delegates from different towns and counties throughout England?' asked Horne Tooke.[22] Eventually, following evidence from Sheridan, Pitt conceded that there had been delegates from county towns and cities – but he managed to avoid the dangerous word 'convention', with its reminder of post-revolutionary France.

Erskine and Horne Tooke called eighteen witnesses in all. One, Major John Cartwright, recalled Horne Tooke, on several occasions, comparing his political views to a stagecoach journey. Whilst others might choose to go all the way to Windsor, he would get out at Hounslow.[23] As a line of defence the analogy might prove Horne Tooke's lack of treasonable intention, but it did so by implying the guilt of others – whose trials were yet to come. Hazlitt disapproved of Horne Tooke's having 'compromised his friends to screen himself', and many shared his distaste.[24]

The examination of witnesses closed on Friday 21 November. On Saturday, Erskine's junior counsel, Vicary Gibbs, addressed the jury in a four-hour speech. The Attorney General, Sir John Scott, at similar length closed the case for the Crown. The Lord Chief Justice summed up and the jury retired. They returned after only eight minutes with their verdict: Not Guilty.

That evening Cline held a victory supper for Horne Tooke, an occasion he would repeat every year after to commemorate his acquittal. The 'agreeable day' to which William referred in his 1803 letter to his mother was probably one of these anniversary celebrations. He met some 'good company' there, but complained that Erskine was 'so full of himself' that he always dominated the conversation 'without any right, either from superior knowledge or understanding'.[25] He was not the only one to find the great lawyer pleased with himself; he was caricatured by Gillray as Councellor Ego.

FIGURE 23 Thomas Erskine, 1st Baron Erskine
(*Councellor Ego – i.e. little I, myself I*), by James Gillray (1798).

As with all Gillray's cartoons, the caricature is full of
satirical detail. Erskine is depicted as a lower-case letter i,
with the dot being a revolutionary cap of liberty.
(BM Satires 9246)

15

PITT'S GAGGING ACTS

Misfortune must make very near approaches to the
great mass of mankind before it excites their alarm.

(William Morgan[1])

After Horne Tooke's acquittal the cases against the remaining prisoners were dropped – all save one, John Thelwall. He was tried in December 1794 and acquitted. For all those tried acquittal came at a cost. There was a strong feeling that they should have been tried for sedition, in which case they would undoubtedly have been found guilty. William Wyndham, Minister at War[2] under Pitt, referred to the defendants as 'acquitted felons'.[3] The trial was over but the name stuck.

Horne Tooke returned to Wimbledon. Prison had taxed him physically and mentally but his garden helped to restore his health and his spirits. He was soon entertaining again and, in the 1796 election, stood as an independent candidate for Westminster. He lost but was eventually persuaded to accept the seat for Old Sarum, one of the most notorious 'rotten boroughs'; with only six men on the electoral role it was effectively the gift of the patron. If this was not irony enough, as soon as he had taken his oath, his right to a seat was challenged on the grounds that he held holy orders. He was allowed to remain until parliament was dissolved in 1802. After that his political energies were devoted to his protégé, Francis Burdett, who became a prominent reformist MP, and was said to be designated by Napoleon as President of England should a successful invasion be achieved.[4]

Horne Tooke's health declined but he continued to enjoy his garden, growing vegetables in addition to gooseberries, currants, strawberries and other fruit. His eldest brother, Benjamin, a successful market gardener who introduced the Saratoga strawberry plant to Britain, was doubtless an early influence and mentor. Horne Tooke wrote, most famously *The Diversions of Purley*, a work in the form of a dialogue on philology but, as well as language and grammar, the discussion explored philosophical and political ideas. He entertained with lavish meals, his deteriorating health not affecting his prodigious appetite. Cline and Erskine were regular visitors and there were new friends, amongst them the poet Samuel Taylor Coleridge, the essayist William Hazlitt, and the young scientist Humphry Davy who greatly impressed Horne Tooke.

As he became more frail Horne Tooke prepared for his death by having a vault installed in his kitchen garden and, beside it, a piece of black Irish marble on which was inscribed:

John Horne Tooke
Late Proprietor, and now occupier of this spot,
was
born in June, 1736
and
Died in the ... Year of His Age,
Content & Grateful.[5]

Horne Tooke's idea was that a summerhouse should be built above the vault, creating a happy place for his daughters to sit and look out across the green.

Horne Tooke died on 18 March 1813 and plans were made, in accordance with his wishes, for his burial without a funeral ceremony. Burdett wrote to Horne Tooke's friends to invite them to attend. Pasted in my autograph book is the letter which William received:

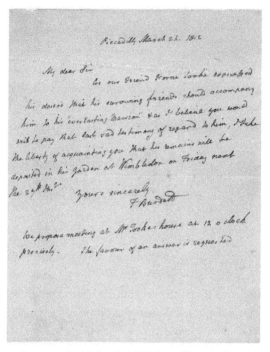

FIGURE 24 Letter of 24 March 1812 from Francis Burdett inviting William Morgan to attend the funeral of John Horne Tooke.

Piccadilly March 21. 1812

My dear Sir

As our friend Horne Tooke expressed his desire that his sur-viving friends should accompany him to his 'everlasting Mansion' & as I believe you would wish to pay that last sad testimony of regard to him, I take the liberty of acquainting you that his remains will be deposited in his garden at Wimbledon on Friday next the 27th Ins^t.

Yours sincerely
F Burdett

We propose meeting at Mr Tooke's house at 12 o'clock precisely. The favour of an answer is requested.

Samuel Rogers received a similar letter and, in this, Burdett writes that Horne Tooke wanted his 'few real friends' to attend.[6] Clearly those invited were held in particular affection by Horne Tooke. But the garden burial never took place. The vault flooded, and anyway the family decided that a tomb in the garden would lower the value of the property; also they wanted a proper Christian service. Instead, he was interred beside his mother and sister in the parish church of Ealing and the funeral service was read 'in a very audible and impressive tone of voice by the Reverend Coulston Carr'.[7] William was there, although, as he told his cousin John Price, he had been ill for most of the year and, at Christmas 1812, considered he was 'still an invalid'.[8]

Rogers is not listed amongst the mourners but the name 'Mr Hardy' appears, presumably Thomas Hardy, whose treason trial preceded Horne Tooke's. Hardy, whose wife and unborn child had died while he was in the Tower, initially returned to shoemaking – with uneven success. He continued to be politically active, though less prominently than before. He died, aged eighty, in 1832 and was buried in Bunhill Fields. As for the radical societies which had excited so much government anxiety, the Society for Promoting Constitutional Information met only once more. The London Corresponding Society, however, survived, its administration taken over by Francis Place, a young journeyman breeches-maker, an autodidact and a prodigious diarist.[9]

William lost none of his reformist energy and scorned 'an intimation' from Pitt 'to the effect that if he would employ his pen on the side of the Government he would "find his account"'.[10] William's incisive mind and business acumen would have given valuable assistance to the struggling economy, but Pitt should have known better. William was not to be bought. Far from it, in 1795 he published a second edition of *A Review of Dr Price's Writings on the Subject of the Finances of this Kingdom*, with a supplement stating the current amount of the national debt.

The year 1795 was appalling. Military campaigns in Europe were disastrous, whilst at home the weather made life miserable. The harvest was all but ruined by the long hot summer of 1794; yields were low and the quality poor. The year began with exceptionally severe frosts followed by heavy snow and then flooding. Throughout the country the bitter cold caused great hardship; food stocks were low and prices high. The needs of the navy and the army put a further strain on resources, and the hungry began to march in protest. In Bury St Edmunds people seized meat from butchers' stalls, and in Devon labourers dressed in skirts to look like housewives, shouted 'We cannot starve', as they wrecked a mill that supplied the fleet.[11] In London an angry mob hissed, 'No Pitt, no War, Bread, Bread, Peace, Peace', as the king was driven in procession from Buckingham Palace to open parliament. Stones were thrown and one broke the window near where George III was sitting.[12] Claiming that the window was shattered by a bullet, Pitt had just the pretext he needed in order to introduce legislation to silence radical voices. It is even possible that the government orchestrated the attack on the king in order to give them a reason for rushing two bills through parliament. The first was the Treason Act, which extended the definition of high treason (the 1794 acquittals still rankled), and the second was the Seditious Meetings Act, which prohibited public gatherings of more than fifty people without a magistrate's licence. They were referred to as the Two Acts – no one needed to have the names spelt out – or, more popularly, Pitt's Gagging Acts.

The year had begun with a frozen Thames, the ice thick enough to support one of the famous Frost Fairs. A picture of a similar fair in 1814 shows children playing skittles and in swing-boats, a whole sheep prepared for roasting, and busy stalls selling chestnuts and 'Good Gin'. It all looks great fun – if you were warmly dressed and well fed.

The Equitable office at Blackfriars was just a short walk from the frozen Thames, but William was not fooled by the festive atmosphere. He was worried about the spiralling cost of the war and the damage it was doing to the country's economy. He felt he had to speak out, and in 1796 he published *Facts Addressed to the Serious Attention of the People of Great Britain respecting the Expence [sic] of the War and the State of the National Debt*. His purpose is clear in his title, but he cannot restrain himself from a passionate outburst in his preface against the 'carnage and miseries' of the war, which ought to 'sicken every friend of humanity'. In a withering reproach he observes that people show little concern when the 'death and desolation' of warfare are far off. 'Misfortune

FIGURE 25 *A View on the River Thames between London and Blackfriars Bridges in the hard Frost 1814,* by George Thompson. *(British Museum Crace 1878 VIII.Frost Fair.11)*

must make very near approaches to the great mass of mankind before it excites their alarm.' His appeal, therefore, is to the 'lower passion of self-interest' and he promises to confine himself to an examination of the effects which the war had already produced on the *finances* of the country, and to expose the 'errors and misconduct of ministers'. William allows that there might be differences of opinion about the justice and necessity of the war, but there could be none about its huge expense.[13]

But there is another comment before he begins his financial analysis. The war is supposedly being fought to defend 'property, social order, and the religion of mankind'. William is cynical. 'Have not wars ... always been found to destroy the property of a nation? and have not the crusades which have hitherto been carried on in the name of religion invariably disgraced and ruined the cause they professed to maintain?' In an ironic footnote he adds, 'This, I believe, is the first crusade for social order.'[14]

Every war, he argues, and there had been many in the eighteenth century, proved more expensive than those which preceded it, the debt arising from the current war being more than double that incurred during a similar period by the war with America only two decades earlier. He presents tables showing detailed comparisons of the expenses of the two wars and, in a waspish comment on the accounts of the navy, the army and the ordnance, he complains that it is as if they are 'proud of keeping pace with each other, and surpassing those that have preceded them, continually increasing in the enormity of their amount'.[15] But his chief target is the government, and not so much for the way in which the public money has been spent, but the manner in which it has been borrowed, and the inadequate measures in place to fund the borrowing. He is dismissive about licences to wear hair powder or kill game, and taxes on 'insignificant articles', which raise very little revenue and cost money to collect. His most chilling warning is a comparison with the financial situation in France which had led to the Revolution. 'New loans became necessary to pay the interest of former loans. The mass of debt continued to accumulate, till at length it overwhelmed public credit and buried the government in its ruins'.[16]

The booklet went into four editions before the end of the year and, in the same year (1796), William published *Additional Facts Addressed to the Serious Attention of the People of Great Britain respecting the Expence of the War and the State of the National Debt*. The arguments and warnings remain the same but more urgent, not least because some of his previous estimates of ongoing expenditure were, he realises, too conservative. Once again the publication went into four editions. William was to write two more booklets in the next four years, each with the aim of alerting the public to the spiralling cost of the war and the precarious state of the nation's finances.

Meanwhile at the Equitable he was dealing with a very different problem: his careful management had built healthy funds for the Society, which led members to expect a bonanza. In March 1793 he gave an address to the General Council of the Equitable in which he warned about the dangers of not thinking ahead; the Society's capital might seem large, but it would be needed for future claims. In a second address, on 3 December 1795, he was more blunt in anticipating the greed of his audience. By this time the capital had reached nearly a million which, he knew, would lead 'the uninformed to entertain very dangerous and extravagant opinions of its opulence'. Would any of his listeners admit to being uninformed? To be on the safe side he rammed home the point: since the capital was likely to increase this would 'not only mislead the ignorant, but probably have the alarming effect of awakening the passions of the covetous and self-interested'.[17]

Surely this would silence his critics, but, no, he came under constant pressure to use the surplus to reduce premiums or distribute bonuses. Eventually his disagreement with the directors became so entrenched that he took himself, together with all his papers, off to 'a stronghold in Wales' where he held out – and won the day.[18] Most probably the stronghold was Bridgend, always dear to William's heart, and the self-imposed exile from London would have been a way of sneaking a visit to his sisters, Kitty and Nancy – and their families.

A letter written by William from Bridgend on 10 September 1796 seems to have been written on a holiday rather than an escape from the Equitable, but it gives a reminder of his distance from London.[19] Planning his return, he writes, 'I mean to begin my journey on Tuesday next and hope to drink tea at Stamford Hill on Sunday afternoon'. Six days' travel, despite much improved roads – by the end of the century Thomas Telford and John McAdam were designing and building sturdy bridges and roads with solid surfaces, and turnpike trusts were making for better maintenance.[20] Stagecoaches and the postal service had become speedier, so William's slow progress suggests that he travelled in his own carriage, allowing time to rest his horses.

The same letter sheds light on his affection for his children. He writes to eight-year-old Susan, whom he addresses as his 'dear little Maid'; he is missing her and her brother, 'dear little William', aged five, whom he left at Stamford Hill with 'a heavy heart'. Only Sarah, twelve years old, had accompanied her father, and he describes an outing to the seaside in a family party, nineteen strong, where Sarah had raced 'so famously' on the sands that she beat her cousins, though they were both of them boys and one year older than her.[21] William tells Susan he wished she had been there to show them that 'they had not yet seen my most nimble little girl', and he promises her walks and seaside bathing the following year.

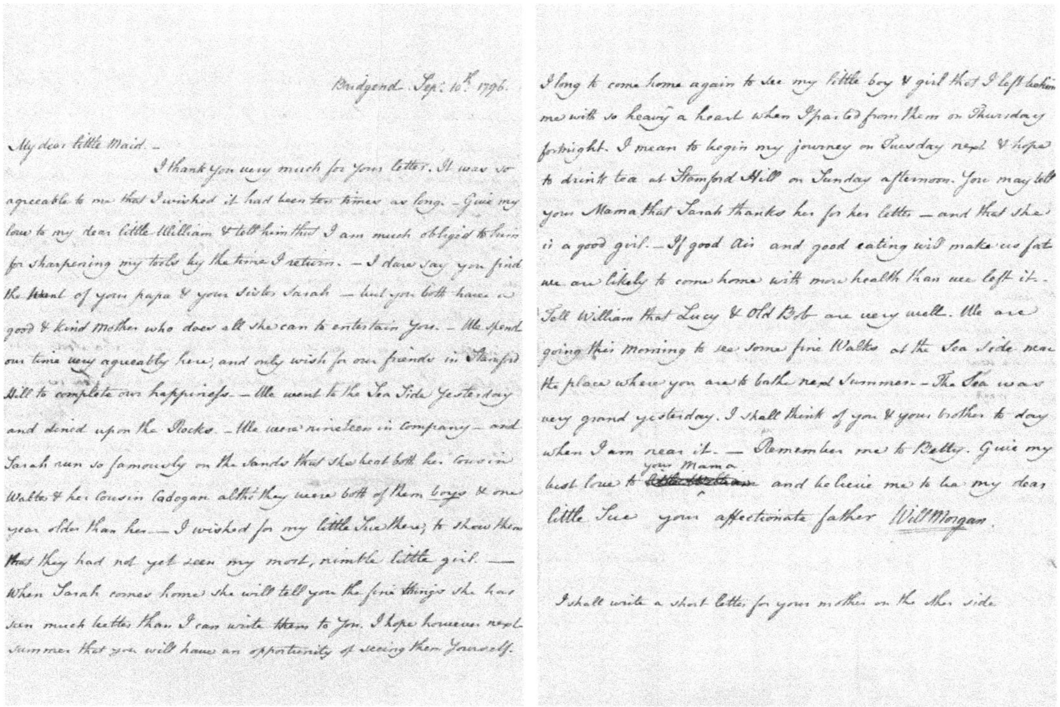

Bridgend Sep: 10th 1796.

My dear little Maid. –

I thank you very much for your letter. It was so agreeable to me that I wished it had been ten times as long. – Give my love to my dear little William & tell him that I am much obliged to him for sharpening my tools by the time I return. – I dare say you find the want of your papa & your sister Sarah – but you both have a good & kind mother who does all she can to entertain you. – We spend our time very agreeably here; and only wish for our friends in Stamford Hill to complete our happiness. – We went to the Sea Side yesterday and dined upon the Rocks. – We were nineteen in company – and Sarah ran so famously on the Sands that she beat both her Cousin Walter & her Cousin Cadogan altho' they were both of them boys & one year older than her. – I wished for my little Sue there, to show them that they had not yet seen my most nimble little girl. – When Sarah comes home she will tell you the fine things she has seen much better than I can write them to you. I hope however next summer that you will have an opportunity of seeing them yourself.

I long to come home again to see my little boy & girl that I left behind me with so heavy a heart when I parted from them on Thursday fortnight. I mean to begin my journey on Tuesday next & hope to drink tea at Stamford Hill on Sunday afternoon. You may tell your Mama that Sarah thanks her for her letter – and that she is a good girl. – If good air and good eating will make us fat we are likely to come home with more health than we left it. – Tell William that Lucy & Old Bob are very well. We are going this morning to see some fine Walks at the Sea Side near the place where you are to bathe next Summer. – The Sea was very grand yesterday. I shall think of you & your brother to day when I am near it. – Remember me to Betty. Give my best love to ~~little William~~ your Mama and believe me to be my dear little Sue your affectionate father Will Morgan

I shall write a short letter for your mother on the other side

Figures 26a and 26b William Morgan's letter of 1796 to his younger daughter Susan (whom he addresses as 'my dear little Maid').

Writing from Bridgend, William tells Susan about a seaside picnic with her Morgan cousins.

The cousins whom Sarah beat in her race on the sands were Nancy's eldest son, Walter, and Kitty's youngest son, Cadogan. Victory over Cadogan may not have been too taxing, for Kitty was an indulgent mother who 'would fain place a pillow under everybody's head'.[22] Beating Walter was quite another matter, for Nancy's children were brought up on tough love. Walter remembered trying to escape return to boarding school by throwing himself off his pony into the mud, in the hope that he would be kept at home. Nancy had dressed him in clean clothes, made him remount and set off again for school.

William's trips to Bridgend sometimes included a visit to Llandough Castle near Cowbridge to see his uncle Samuel Price and, after Samuel's death in 1777, to see his widow, Catherine. As well as supplying Catherine with opium, William managed some of her investments, writing regularly to keep her informed about her financial affairs. Catherine, for her part, sent presents of game from the country. William's thanks for her gifts could be matter-of-fact to the point of bluntness. 'I thank you for the present of hares which you sent me', he wrote on one occasion, adding 'and though they were both so putrid as to force us to bury them immediately, I do not feel the less obliged by your kind intention'.[23]

After Catherine's death in 1806 William managed investments for her son, John, and also found supplies of good Kentish hops for him. John, like his mother, sent frequent presents of game – always a turkey at Christmas and often woodcock as well. William's letters to his cousin give telling glimpses of his worries about the Equitable and about the country.

INVASION, PANIC
AND MUTINY

Our situation becomes daily more perilous, and unless a more
temperate and frugal policy be soon adopted, and those abuses
be reformed which have proved the source of all our misfortunes,
it will be as vain to cherish the hope of avoiding destruction as
it has hitherto been unprofitable to bewail the progress of it.

(William Morgan[1])

William's anxieties about the national situation increased during
the last decade of the eighteenth century – and with good reason. The country was at war and threatened with invasion. A cartoon
by Isaac Cruikshank published in 1798 encapsulates the fears and the
rumours. *Intending Bonne Farte raising a Southerly Wind* shows hot-
air balloons, parachutes, weapons, tents, wagons, even a tiny guillotine,
being blown towards England as a bare-bottomed Napoleon breaks
wind on the French coast. At sea an immense raft and a huge sea
monster carry French troops across the Channel, whilst, on the cliffs
of Dover, Sheridan and Fox, each wearing the bonnet-rouge of liberty,
wait to welcome them, saying, 'How fragrant is this Southerly Breeze',
and at their feet sits William's friend, Horne Tooke. The subtitle, *Or a
Sketch of the Intended Invincible Invasion found at the door of Brooks's in
St James's Street*, underlines the current belief that many frequenters of
Brooks's club were supporters of Napoleon and the French Republic.

The scatological humour may be puerile, but the fears were real
and had some foundation. In late 1796 the French had planned a

FIGURE 27 *Intending Bonne Farte raising a Southerly Wind,*
by Isaac Cruikshank (1798).
(BM Satires 9172)

three-pronged invasion, the main part of which with some 15,000 troops
was to land in Bantry Bay in Ireland.[2] Two smaller attacks (effectively
diversionary tactics) were planned for Newcastle in the north-east and
Bristol in the south-west. In the event, the Irish 'prong' was thwarted by
fierce storms, which scattered the ships and drowned many thousands.
Bad weather prevented the Newcastle attack and nearly did for the
third and final attempt, which, instead of taking Bristol, was blown off
course and eventually landed at Fishguard in Pembrokeshire. But this
was a small force, consisting mainly of conscripted ex-convicts, many of
whom lost no time in deserting and looting local farms. In the face of
militia and volunteers (and in the mistaken belief that a crowd of local
women wrapped in red flannel cloaks was a regiment of redcoats) the
remaining men quickly surrendered.[3]

The Fishguard landing was more a French fiasco than a British
victory, but it showed that invasion was a distinct possibility and it sent
a shudder of fear through the nation. When government agents began
making inventories of stock in coastal regions, farmers suspected that

their corn and cattle might be requisitioned and they hurried to market, selling even at reduced prices. Panic spread quickly. People began to withdraw money from the country banks leading to a run on the Bank of England, where by the end of February 1797 the stock of bullion fell to a record low of £1.2 million, with over £100,000 being withdrawn every day. If the situation continued the nation would be bankrupt.[4]

Pitt persuaded the king not only to agree to a special meeting of the Privy Council at Buckingham Palace but also to hold it on 26 February 1797 – a Sunday. The emergency was addressed by an Order in Council allowing the suspension of cash payments and the issue of notes without the promise to back them up with gold: paper money.[5]

William was appalled and declared, 'The dignity of British credit has, in a moment, been reduced from its lofty pre-eminence to a state of the most humiliating degradation.'[6] A cartoon by Gillray (clearly not in the pay of the government at the time) shows similar dismay. Pitt, thin but with his stomach engorged with gold coins, bestrides the rotunda of the Bank of England, on which he defecates and vomits paper money. Around his neck is a chain and padlock labelled 'Power of Securing Public Goods', in his hand the 'Key of Public Property', whilst on his head he has asses' ears. The picture and the title invert the Midas myth – here is Pitt turning gold to paper – and the satirical detail includes Pitt's Foxite opponents whispering, 'Midas has ears', and his allies belittled by being depicted as naked children (technically putti).[7]

In April 1797, a month after Gillray's cartoon appeared, William published *An Appeal to the People of Great Britain*.[8] Much of it repeated the outrage of *Facts Addressed to the Serious Attention of the People of Great Britain*,[9] which he had published the previous year. The 'Welsh temper' of his youth is still as fiery. He feels indignation at money squandered in 'ruinous prodigality'. He is scathing about 'the extravagance of exorbitant expenses', and he castigates Pitt and his ministers for their incompetence. Once again he reveals the costs, and hidden costs, of the war providing comparative tables to support his arguments. Again, he deplores in particular the extraordinary expenses – those which had been incurred without prior consent from parliament and which had risen at an alarming rate. Being 'extraordinary' these could be left out of

FIGURE 28 *Midas, Transmuting all into ~~Gold~~ Paper*,
by James Gillray (1797).
(BM Satires 8995)

statements of expenses, so that the public were unaware of the gravity
of the financial situation.

He illustrates his argument by focusing on the 'long list' of army
extraordinaries and points out that they included 'the travelling expences
[*sic*] of confidential emissaries, the salaries of civil officers, and other
articles which seem to have very little connection with the army'.[10]

As for the colonial governors, they had made claims amounting to £1,364,806, the lion's share going to the governor of St Domingo.[11] If, asks an exasperated William, 'the possession of a narrow neck of land has obliged the governor of that district to expend above £1,100,000 ... what must have been the amount of that expenditure if the whole island had been in our possession?'[12]

Nor were the governors alone in claiming expenses. Similar discretionary powers were afforded to military commanders, the commissaries, the deputy commissaries, the deputy-paymasters and 'almost every other officer who is employed in public service'.[13]

The navy debt was, if anything, worse, since the naval extraordinaries were not itemised but presented as a general amount. Figures set before parliament were inaccurate or fudged, so that the true magnitude of the national debt was concealed. As well as the spending on army and navy, the war triggered other expenses, in the form of loans and subsidies to foreign powers in order to secure their alliance. With the Austrian Emperor being furnished with 'almost unbounded credit', William saw the situation as 'ruinous'.[14] Nor would it be immediately improved by the ending of hostilities:

> The expenses necessarily attending the conclusion of every war, and which arise from the continuance of pay to the forces, the calling home of the troops, and the arrears in the different departments, etc. will probably amount to half as much as the expenditure of another campaign.[15]

In his concluding chapters William attacks the directors of the Bank of England for their failure to stand up to the Treasury demands for more and more money to fund the war. He is particularly bitter about the 'fictitious coinage of paper', which, far from providing a solution, gives the nation 'a very false appearance of wealth and magnificence'.[16] His judgement on the Bank and the directors is severe: 'I am no friend to institutions so grossly perverted and abused. Too long have they served as arsenals from which avarice and ambition have been furnished with implements to deluge the world with blood.'[17]

The publication went into four editions, and William Morgan was hailed at a meeting of the London Corresponding Society as a 'great calculator'. Calculator can be something of a weasel word but there is no suggestion of scheming in this epithet; it is a compliment and William's publication is recommended for 'serious perusal'.[18] And others took note. Was it William's analysis which prompted Pitt's government in 1800 to pass an Act of Indemnity protecting themselves against any legal claims arising from the handling of government finances in the war years?[19]

And then a different threat challenged the country's security. Anger about pay and conditions was simmering in the navy. At Spithead on 15 April 1797 sailors refused the command to weigh anchor and sail, instead giving three cheers of defiance. Mutiny. The crisis should have been short-lived; the men's demands were reasonable: fairer pay (they had not had any increase since 1658), a share in prize money, and improved care for the sick and injured. The government recognised that it needed to make concessions, and a bill to increase seamen's pay was rushed through parliament. Poor communications, however, delayed the information reaching the men, and the mutiny spread – most seriously to Nore, a sandbank and anchorage on the Thames estuary, which was a strategic holding area for vessels entering and leaving the Thames. For William at Blackfriars and other Londoners this meant that the city was vulnerable either to invasion or to a blockade, as the rebels began to seize food and water from passing merchant ships. Pitt held his nerve and introduced two emergency bills, imposing penalties on anyone encouraging sedition in the forces or communicating with the mutineers. Gradually the men gave in. The mutiny was over, but it had been a tense time and left a shadow of disquiet in the country.

Meanwhile the Irish were chafing under British rule and, whilst hoping for help from the French, began a series of bloody skirmishes. Reports by spies so alarmed the government that, early in 1798, martial law was imposed. There was violence and brutality on both sides. George wrote to his mother about what he had heard:

> roads covered with putrid bodies . . . burning and burnt houses . . .
> villages changed into barracks and in many places daily executions.

Some flogged, some half hanged, and the horrid spear spikes elevated with the bleeding heads of the rebels fixed on them.[20]

The rebellion was quashed by the end of 1798, but not before the French had managed to land a small contingent in County Mayo, creating fresh anxiety about a mainland invasion and renewed suspicion about dissidents. In 1800 the Act of Union with Ireland was passed, although without Catholic emancipation, a sticking point which contributed to Pitt's resignation in 1801.

THE REIGN OF TERROR

We appear to be like the infatuated mariners of old who in the
midst of shoals and quick sands suffered themselves to be allured
by the song of the Siren, nor awoke from their delusion until
they were ingulphed [*sic*] in the waves that heaved around them.

(William Morgan[1])

Anxiety about the safety of the nation and its precarious finances
affected the government's ability to raise money. Government
stocks slipped to half their price, making William gloomy at a domestic as well as a national level. When his brother George asked how he
should invest £500, William's advice was cynical: 'bury it'.[2] Others were
bold enough to treat the unpopular funds as an investment opportunity.
With government stocks at a bargain price, one wealthy merchant in
Hull, Thomas Thompson, decided that 'if the French landed it would
make no difference whether he met his fate as a rich man or a poor
one'.[3] He invested heavily in the funds and pocketed a large profit when
the danger had passed and the prices rose.

Neither brother would have approved of financial gains made
through such speculation. In all his addresses to the Equitable William
warned against the risks of ruin through greedy investment. George had
moral qualms about material wealth. 'Make a man rich', he declared,

and you make him lazy, luxurious and the slave of his own passions . . . the slave of the times, the interested friend of every established abuse, and the most obstinate supporter of all religious and

civil tyranny. Make a man rich and you expose him to all the arts, the intrigues, the flatteries, the delusions, and all the poisonous frauds by which the worst characters in society rise on the idleness and passions of the most foolish.[4]

The two brothers lived near enough to see each other regularly and the family bonds were strong. When George looked for a larger house to accommodate his growing family and his private pupils, he explained in a letter to his mother that he wanted to find one where he could continue 'to be frequently visiting you and my brother'.[5] Their political views were similar and the Sunday evening gatherings at Stamford Hill were as important to George as they were to William.

Science was opening a new world of discovery, and William and George enjoyed the exchange of ideas in the scientific community. In his 1785 'X-ray' paper William acknowledges a debt to a close friend of George's, 'the ingenious Mr Brook of Norwich'. Brook's advice on the method of boiling mercury had been the 'chief cause of [his] success in these experiments'.[6]

But above all both William and George enjoyed watching their children grow up. By the end of 1797 William's daughters, Sarah and Susan, were thirteen and nine, his son, William, was six and the baby, John, was nearly a year old. George's daughter, also Sarah, was thirteen and his seven sons ranged in age from eleven years to six months. William's 1796 letter shows his interest in their upbringing. Sarah's prowess on the sands and his description of Susan as his 'most nimble little girl' suggest that the girls enjoyed tomboy freedom in the spacious gardens at Stamford Hill – and with their father's blessing. In the same letter he sends his love to five-year-old William and thanks him for sharpening his tools (supervised, no doubt, but giving trust and responsibility to the little boy). The message speaks of a close bond between father and son as well as anticipating Maria Edgeworth's 1798 work *Practical Education*, with its emphasis on fresh air, play and what today we call hands-on learning.

Both brothers would have been familiar with Mary Wollstonecraft's *Thoughts on the Education of Daughters*, which was written in 1787 when she was running her small school at Newington Green, a time when Richard Price was something of a father figure to her and a time when a woman was expected merely to 'know just enough arithmetic to do household accounts and just enough geography to converse with her husband and friends'.[7] Wollstonecraft and the Morgans wanted more for their daughters and, despite the depressing national and international situation, both the young Morgan families seem to have been enjoying a carefree childhood. It was not to last.

In the summer of 1798, shortly after his youngest son, Septimus, had celebrated his first birthday, George began to look unwell. No one was unduly alarmed. George was athletic and strong; he walked, he fenced, he played cricket, and later that summer he swam as usual on a visit to Southerndown. On his return to London, just a couple of weeks after his sixth son's third birthday, George was 'seized with a rigor, nausea, and other symptoms of fever'.[8] Doctors were called and William came to look after his brother; he nursed him day and night. The sickroom of a much-loved family member must have evoked memories of Richard Price's dying days. But George was young; surely this was different.

He died on Saturday 17 November 1798, two days before another son's fifth birthday. George was forty-four; he left a family of eight young children and his widow, Ann, two months pregnant. The family blamed his death on his chemical experiments, during which George was thought to have inhaled poisonous fumes,[9] but in his obituary his final illness was described as 'pulmonary consumption'.[10] Whatever the cause, his death was a shock and its consequences were to be far-reaching.

William felt the loss very keenly. George was his only surviving brother and they had been very close. George's son, writing some years later, claimed that, after his father's death, the gatherings at Stamford Hill lost their passion: 'the moving spirit was no longer present, the survivor was too much of a conservative and too timid to suit these radical politicians'.[11] George may have been the more fiery of the brothers but William was as committed to the reformist movement. If the Sunday evenings had lost some of their fervour, William's grief must be part of the reason. But not all. The 1790s have become known as Pitt's Reign of Terror; the Gagging Acts of 1795 had suppressed the reformist movement and, by the turn of the century, radical voices had been muzzled in the atmosphere of suspicion and spying – not unlike the McCarthyism of 1950s America.

At least one of William's guests, Thomas Paine, had fled the country. Others chose to avoid political activity. Horne Tooke and the rest of the twelve apostles had been acquitted in 1794 but theirs were not the only cases tried. There were more trials for treason and sedition in the 1790s in Great Britain than ever before or after in its history.[12] Samuel Rogers was caught up in one of these in 1796 when he was called as a witness in the trial for high treason of William Stone, a wealthy coal merchant of Newington Green. Stone's correspondence with his brother in France led government spies to suspect him of involvement in an invasion plot. Stone, defended by Erskine, was acquitted (to claps and huzzas in the courtroom), but afterwards Rogers 'was less inclined than ever to take any active part in the political agitations of a time when the Habeas Corpus Act was suspended and there was a reign of terror in England'. Rogers discontinued his diary with its account of conversations at the dinner table – toxic material, his biographer suggests, since the 'brief record of the toasts drunk at Mr. Morgan's table, and of Paine's remarks on kings might have been made evidence against him had the Ministers of the day known that it was in existence'.[13] Rogers's diary entry would, undoubtedly, have incriminated William.

By the turn of the century the carnage of the French Revolution had led many to revise their enthusiastic welcome of the uprising in 1789. Or did Pitt's Gagging Acts make it expedient to do so? A few were bold enough to applaud the principles of the Revolution although deploring its atrocities. William was one of them, but he had to be careful. He was doubtless shocked by the fate of one of his contemporaries, Gilbert Wakefield.

Wakefield's 'crime' was to publish in 1798 *A Reply to the Bishop of Llandaff*. The Bishop of Llandaff (at one time a friend of William's)[14] was something of a renegade. Originally liberal, he had changed his opinions and written a pamphlet attacking the French Revolution and supporting the war and Pitt's taxes to finance the war. Wakefield shared William's views about the ruinous cost of the war but in his *Reply* he focused not only on the economy but also on Pitt and his foreign secretary, Grenville, calling them 'self-opinionated, arrogant, defamatory, and despotical'. He went further accusing them of having 'engendered sham plots, false alarms, and visionary assassinations, for the purposes of deluding the unwary, [to] establish a military despotism in England'.[15]

Both Wakefield and his publisher, Joseph Johnson, were arrested. Johnson was prosecuted for publishing seditious literature, fined £50 and sentenced to three months in the King's Bench prison. Wakefield was found guilty of seditious libel and was sentenced to two years in the notoriously harsh Dorchester jail. He died in 1801, three months after his release, of typhus contracted in prison.

But Wakefield's was not the only reply to the Bishop of Llandaff. The poets Blake and Wordsworth both wrote in similar vein. Why were they not arrested?

They did not publish their replies. It is highly likely that William knew about Wordsworth's unpublished pamphlet. He and Wordsworth had a mutual friend in Samuel Rogers. (Rogers even lent Wordsworth his court suit when he received his laureateship – a tight fit, as Wordsworth was the larger of the two men.[16]) The imprisonment of Wakefield and Johnson were matters of grave concern within the circle. Enforced censorship made everyone nervous. George, shortly before his death, had planned to write a biography of Richard Price but he was deterred by the possibility of arrest. 'I am sadly frightened by poor Johnstone's [*sic*] conviction. Most probably he will be kept in gaol long enough to ruin both his health and property.'[17]

The relentless pressure on the reformists intensified with the publication of the *Anti-Jacobin*,[18] which appeared every Monday during the parliamentary session of 1797–8, and – by means of some imaginative calculations – claimed a total readership of 50,000.[19] Founded by George Canning, undersecretary for foreign affairs, the paper was blatant government propaganda. Alongside rousing verses and satirical poetry, it offered readers reassurance about the progress of the war and justification of the taxes imposed to finance it. Each issue had three sections, headed 'Lies', 'Misrepresentations' and 'Mistakes', which sought to refute the reports in the *Morning Chronicle* and other opposition papers. With twenty-first-century hindsight it is easy to be sceptical about the assertions and the selective presentation of finances, but, for a nation horrified by the events in France and fearful of invasion, they were probably convincing and the patriotic songs must have stirred faltering hearts:

> Let France in savage accents sing
>   Her bloody Revolution;
> We prize our country, love our King,
>   Adore our Constitution;
> For these we'll every danger face,
>   And quit our rustic labours;
> Our ploughs to firelocks shall give place,
>   Our scythes be chang'd to sabres.
> And, clad in arms, our Song shall be,
>   'O give us Death – or Victory!'[20]

William was incensed and frustrated. He could see through the government's glib assurances; he had already tried in three publications to alert the public to the dangerous state of the economy but with no useful result. Now Wakefield's imprisonment, Wordsworth's decision to withhold his pamphlet, and his late brother's fear of publishing a biography of their uncle were a warning to him. Despite the obvious risks he published in 1801 *A Comparative View of the Public Finances from the Beginning to the Close of the Late Administration.* As in previous publications he was careful to insist that he had 'wholly abstained from every discussion which was not immediately connected with the public finances'.[21] But he slips in sideswipes wherever he can, not least at Pitt, who had resigned office in 1801. There is a touch of glee in his comment that 'the friends of a fallen minister are seldom very strenuous in his support', and heavy irony in his adding that 'the negligent deficiencies of his friends have, however, been abundantly supplied by the glowing account which he has given of his own conduct'.[22]

Once again William provides figures and analysis to show the size of the national debt and its rate of increase – much more rapid than in previous wars. He is critical of the ways in which money has been raised, in particular the introduction of paper money and the false sense of wealth it has created: 'a new mine has been laid open, where millions may be coined in a few hours, and the loans which lately were deemed impracticable, may be raised with as little difficulty as they have been voted'.[23] He deplores the tax burden which the government has placed on future generations and is scathing about Pitt's self-congratulation on leaving office. He finishes on a note of deep gloom:

> As a sincere friend to the liberty and happiness of Great Britain, I cannot contemplate its present situation and future prospects without concern and dismay . . . We appear to be like the infatuated mariners of old, who in the midst of shoals and quicksands suffered themselves to be allured by the song of the Siren, nor awoke from their delusion until they were ingulphed in the waves that heaved around them.[24]

The war was to last for twelve more expensive years, but Britain was not reduced to bankruptcy. So was William wrong to be so gloomy?

Not wrong, but of his time. In the twenty-first century public debt is a fact of economic life – viewed in the context of Gross Domestic Product, and measured alongside inflation and economic growth. For eighteenth-century politicians and commentators the national debt was something new and nasty, something to be got rid of as soon as possible, and William's was by no means a lone voice. The philosopher David Hume, earlier in the century, wrote that 'the endless increase of national debt is the direct road to national ruin'.[25] Similarly Adam Smith thought that any great increase in public debt would lead to bankruptcy.[26] Looking back we can see what eighteenth-century economists could not anticipate – that the Industrial Revolution saved the situation by generating trade and creating wealth.

18

DIASPORA

We hold these truths to be self-evident, that all men are
created equal, that they are endowed by their Creator
with certain unalienable Rights, that among these
are Life, Liberty and the pursuit of Happiness.
(Thomas Jefferson, *Declaration of Independence*)

George's death left his widow, Ann, managing their home in
Southgate, a house spacious enough to accommodate some of
George's pupils, and where his study alone, doubling as a schoolroom,
measured some sixty feet by twenty-five. Clearly this was more than
Ann needed, even though she had a large family: not only her own
eight children (and a ninth expected), but five others for whom she
had care. Three were the children of a former pupil and paying guest,
William Ashburner, who lived and worked in India – and died there
only two months before George's death; and two were the children of
William Ashburner's sister, Grace Boddington. These two, aged only
five and two, were the innocent victims at the centre of an affair which
scandalised society.

It was quite common for children to be brought up with relatives
or family contacts. Fanny Price in *Mansfield Park* is a fictional example,
and Jane Austen had real-life experience of the practice. Her brother
Edward at the age of fourteen went to live with his wealthy uncle
and aunt, Thomas and Catherine Knight, eventually being adopted
and inheriting their estate.[1] No one thought it irregular when a man
took a mistress, or when he acknowledged his illegitimate children.

Horne Tooke paid for his daughters' education and they came to live with him at Wimbledon. But, when a woman behaved unconventionally, it was another matter. William Godwin's 1798 memoir of Mary Wollstonecraft with its revelations of their free love shocked the respectable world. Grace Boddington broke all the accepted rules. She left her rich husband and comfortable home; she eloped with her husband's cousin; she abandoned her two young children.

In the high-profile court case which followed her flight, Thomas Erskine was counsel for Samuel Boddington, her deserted husband, who sued his cousin for £50,000. Erskine made much of Grace's beauty and hitherto exemplary character. The blame, he argued, lay entirely with her lover, Benjamin Boddington (who anyway pleaded guilty). The jury directed the twenty-four-year-old Benjamin to pay £10,000 damages. He was financially ruined.

Samuel Boddington, having divorced his wife, published *Treachery and Adultery, £10,000 Damages!* giving a full account of the trial. It seems a very public way to deal with a broken heart and wounded pride, but perhaps it is an indication of the level of gossip the case aroused. As for the children, they belonged to a class accustomed to nursery maids and other servants. Nevertheless, Erskine had made much of their being 'deserted, deprived of all the comforts which they had a right to receive from a tender mother'.[2] In Ann Morgan's household they would at least have plenty of young companions to distract them from their loss.

Ann took an 'old-fashioned house' at Stamford Hill. She was, however, hardly downsizing; there were 'numerous rooms, and extensive playgrounds with a large garden', but the rent was moderate and she was nearer to her Morgan in-laws.[3] Here her son, Henry Octavius, was born in June 1799. William's third son, Cadogan, had been born in 1798. There is no record of the month but, whenever it was, both little boys started life at a time of great sadness.

William tried to support her, giving what practical help he could. Fourteen-year-old Sarah was gathered in to be educated with his own daughters, and the boys spent much of their time with their cousins. Ann's granddaughter, however, in a memoir written some years later, suggests that relations between the two households were strained.

William, she felt, did not get on with his sister-in-law and disapproved of her 'easy-going indulgent ways to everyone who belonged to her'.[4] Ann, the daughter of a Norfolk merchant, prosperous from trade with Russia, had been brought up in comfort and luxury. Marriage to a poor clergyman should have taught her to be thrifty and careful, but she was cushioned by gifts from her father and it was her nature to be impulsive and generous. William and Susanna had accrued their wealth slowly and by the same rules of prudent management as he propounded at the Equitable; they struggled to understand Ann's way of life.

After three years at Stamford Hill, Ann took a house at Craven Hill in Bayswater, a move which distanced her from her in-laws and their influence. On the face of it there were other reasons for her decision, chiefly Samuel Boddington's persuasiveness. He wanted to see more of his two children who were still in Ann's care. Craven Hill was closer to his residence in Brook Street near Hyde Park. There were benefits for Sarah, Ann's eldest child and only daughter, who already showed considerable artistic promise. At Craven Hill not only was she nearer to the Royal Academy but she had the opportunity to visit several artists of the time. Better still, Richard Price's old friend, Lord Lansdowne, allowed Sarah free access to his palatial house in Berkeley Square where she was able to study and copy works in his collection of old master paintings.

Life for Ann and her family seemed settled and reasonably secure. Then in 1804 a visit from India by George's erstwhile pupil, Luke Ashburner, changed everything. Recently widowed, Luke wanted to see his family and old friends. Three of his late brother's children, now his wards, were in the care of Ann Morgan, as were the children of his infamous sister, Grace Boddington. After only a very short time, Luke had proposed marriage to Ann's only daughter, Sarah. Luke was thirty-two, twelve years older than Sarah – not such a very big age difference – and from the same social milieu but, fond as she was of Luke, Ann was devastated. The thought of her beloved daughter moving to India was almost unbearable; she gave her consent with a heavy heart and only after Luke had inspired her with his grand plan.

Luke suggested that Ann and her sons should emigrate to America. He would return to India with Sarah and, as soon as he had settled his affairs, they also would emigrate to America. Luke fed Ann a heady mix of ideas about the life they could expect, and she must have heard echoes of her husband as Luke described a land of opportunity, where those who worked hard would thrive and prosper. He promised a society with no blight of aristocracy, a community where enterprise would earn just rewards.

Ann's granddaughter, in her memoir, claims that William was angry and indignant when he heard the news. It is not hard to believe. True, it had been George's dream, but his death had altered everything. How could a young widow with very little money, with a young family ranging in age from teenager to toddler, with no real knowledge of the New World, how could she take such a ridiculous risk? He was not alone in raising objections to Ann's plan; her sisters and other family members, as well as most of her friends, were vehemently opposed to the project.

Despite all the warnings, Ann stuck to her plan, though it was four years before she put it into action. Meanwhile her daughter, Sarah, married Luke Ashburner in 1804. The couple set out for India with the promise that they would eventually emigrate to America. The thought of reunion with her daughter strengthened Ann's resolve and in July 1808 she sailed for Philadelphia in a small American merchant ship, the *Anne Elizabeth*, taking five of her sons with her. Two sons had gone ahead of her; only one son remained in England. The family, together with their cook and her two-year-old daughter, were to share a small cabin for the forty-seven-day journey, buffeted by gales and suffering seasickness as well as the obvious discomforts of living in such a confined space.[5]

Life, when they arrived in America, proved very tough and Ann had no idea how to manage her money. Worse, her beloved daughter, Sarah, never managed to join her in the New World. She and Luke remained in India for thirteen years until 1817, when Luke went ahead to Stockbridge, Massachusetts, leaving Sarah and their children to follow. But the climate – and doubtless the births of five children – had taken a heavy toll on Sarah's health. Instead of sailing to America she returned to England and took a house at Stoke Newington, near to Stamford Hill.

William and Susanna must have been horrified at the change in the charming young woman of whom they were so fond, by this time 'a melancholy wreck of herself'.[6] Ann Morgan, intrepid as ever, immediately made the Atlantic crossing to be with her daughter. In due course Ann took Sarah to Wales, where they stayed at Llandaff Court with William's sister, Nancy Coffin, widowed by now and living with her elder son, Walter, and her daughter, Mary. But the Welsh air and the kindness of her relatives were not enough, and Sarah's health declined. Her husband, Luke, arrived from America but only in time for her final months. She died, aged thirty-six, on 3 January 1820 and was buried in the graveyard of Llandaff Cathedral.

Ann returned once more to England in 1827 at the age of sixty-four and then remained in America until her death in Hudson, New York, aged eighty-three. We can only guess how much she admitted to William and Susanna on those visits, and in letters, about her dwindling funds and the hardships of her life. We do know that all branches of the Morgan families kept in touch and that the cousins from America were always welcomed at Stamford Hill and Bridgend. Sarah in India had corresponded regularly with her cousin, Susan (William's daughter), describing her life in Bombay. Sadly, no letters survive.

THE PRICE OF SUCCESS

I know that nothing depresses the spirits
more than sleepless nights.
(William Morgan[1])

Concern about the cost of the war and anxiety about his brother's family clouded the backdrop to William's life at the Equitable, where work was relentless. He was, as well, faced with a familiar paradox: although the nation's finances were precarious, the Society, thanks to William's prudent management, was prosperous. The members, however, far from being grateful, were restless. Even those who were comfortably off were directly affected by the cost of the war. In his December 1798 budget Pitt proposed a new measure to raise much needed revenue: income tax. It became law in 1799 – a graduated tax beginning with a levy of 2d in the pound on annual incomes of over £60 (less than 1%) and rising on a sliding scale to 2s. (10%) on incomes over £200. Pitt promised it would be only a temporary tax; it was abolished briefly between 1801 and 1803, and again in 1816, but reintroduced in 1841 by Sir Robert Peel. Both Gladstone and Disraeli hoped to remove it but failed to do so. It has become a fact of life and perhaps we can understand why the members of the Equitable wanted a more immediate share in the Society's wealth.

William recognised that the Equitable's prosperity and in particular its surplus raised special problems. Distributing the money amongst members, however attractive in the short term, was liable to leave the Society's finances in a precarious position. On the other hand retaining the surplus or delaying distribution had an inbuilt unfairness for older members, who were likely to die before they received their share. Another problem was that a society cushioned by healthy funds was likely to attract new members. Whilst these newcomers were needed for continuity, was it fair that they should enjoy a share in the surplus accumulated by older members?

William worried; members grumbled. Surely, they argued, the premiums could be reduced or most of the surplus distributed. 'Not safe, equitable or advantageous',[2] said William. Older members whose premiums had helped to create the Society's wealth would be far better served by an increase in their assurances. As for a distribution of the surplus, not before an investigation into the real state of the Equitable's finances. The decision to undertake the investigation was made on Christmas Eve 1798 – shortly after Pitt's income tax budget and little more than a month after George's death. Perhaps the work helped to distract him from his grief; William had over 20,000 calculations to make, a labour which was to take him over two years.[3]

Meanwhile the century ended with wages falling, costs rising and miserable weather. Bitter cold lasted until April, and even in May snowdrifts remained in the fields. There were frosts in June followed by floods in July; crops suffered and food shortages led to protests – conditions which would strengthen William's resolve to be careful with the Equitable's funds. In April 1800, his investigation completed, he gave his third address to the Society. Aged fifty, he had by then served as their Actuary for twenty-five years; he had experience and a successful track record; his words had some clout and he spoke with passion. He reminded members of the 'great error of considering the accumulation of [the Society's] capital as an unequivocal proof of its prosperity'.[4] The nature of life assurance (in the form of the shape of the mortality table) is that there is inevitably a build-up of capital before the claims increase and start diminishing the capital. Moreover, the mortality experiences of each year are likely to vary. Epidemics, warfare and even chance could slew the number of deaths and subsequent claims in any one year.

'Can anything be more absurd, or betray greater ignorance,' he asked, 'than to propose an annual profit and loss account in a concern of this kind; or to regulate the dividend . . . by the success or failure of each year?'[5] He proposed three by-laws: a) that a valuation should be made every ten years; b) that no bonus should be given without a valuation; c) that the value of these bonuses should not exceed two-thirds of the surplus. They were adopted, a measure of the cogency of his arguments and his powers of persuasion – and of his intellect. William had no manual to guide him; he was devising the rules as he went along. Not until 1848, nearly half a century later, was a professional body formed, and not until 1884 was the Institute of Actuaries granted a charter confirming its role and its right to confer qualifications. Looking back in 1933 the President of the Institute, Sir William Elderton, in his tribute in the *Times* praised, as 'his most brilliant performance', William's work on devising a fair method of distributing bonuses.[6]

However fair the method, grumbles were to re-emerge at regular intervals. In 1807 William admitted in a letter to his cousin John Price that the noise and bustle of one of the General Courts had left his head 'hardly in a state to express [him]self intelligibly'.[7] He must have had an even greater headache when in 1808 there was a fresh attempt by some members to have early access to their share of the surplus – an application which was defeated only on a point of order. By the time of the ten-year valuation, completed at the end of 1809, the Society's surplus was £1,615,940, an irresistible temptation to some members, who clamoured for their portion of the pie. As before the most vociferous were the very old, who wanted their share before they died, and the new members who did not want any delay in enjoying the benefit of their investment.

In his address to the General Court on 7 December 1809, William first flattered his listeners: 'Happily the liberality and the good sense of the great majority have hitherto succeeded in quieting the fears of the one, and checking the impatience of the other, and thus in promoting ultimately the real interests of both.' He was, of course, congratulating them on the very restraints he advised. Then he rebuked those who had been 'induced by the great wealth and prosperity of the Society to assure their lives with most extravagant hopes of benefit to themselves and their families'.[8] Who would want to admit to such unintelligent greed? Even with William's recommended caution the bonuses were substantial – an increase of £2 10s. per £100 assured for each year's premium prior to 1 January 1810. For anyone who had taken out a policy in 1770, £100 (40 × £2 10s.) would be added to each £100 assured. In total each £100 originally assured in 1770 had become £390 and would continue to rise.

The members accepted William's recommendations and recorded their thanks for his

> very able and perspicuous report . . . and in particular for the forcible manner in which he has recommended a steady perseverance in the wise, temperate and prudent line of conduct to which the present flourishing state of the Society has been principally owing and which is now more than ever necessary from the increased magnitude of its concerns.[9]

William was too exhausted to take pleasure in his success. 'I would not undergo the fatigue and anxiety of the past year a second time for twice the sum', he told John Price. The letter has a bitter edge. The Equitable's prosperity, he wrote, 'serves only to increase my trouble, and were I to sacrifice my life to it, perhaps I might be mentioned with regret for a few months but, believe me, the great mass would forget me and my services before my family had put off their mourning garments'.[10]

If avarice and argument troubled William, so too did individual problems. The rules and regulations might ensure the profitability of the Equitable but behind every claim lay a human story and there were always difficult cases to consider. Joseph Holl, for example, who died on the day his premium was due. His brother was told that the premium must be paid but he did not do this soon enough, subsequently pleading 'entire ignorance' of the regulations. A compassionate resolution was found: the Court of Directors decided that his claim was ineligible, but the money was paid as a gratuity.[11]

A similar case was that of Mary Price, a maidservant, who had assured £300 on the life of Edmund Phelips but, being out of town, had to rely on a friend to pay the premium for her. One year the friend forgot the payment and Phelips died three days after the expiry of the days of grace. Mary Price pleaded that in her situation and at the age of sixty-six, the loss would be 'sensibly felt'. Compassionate again, the Court allowed her an allowance – as a free donation – of £20 a year during her life.[12]

Not all hard-luck stories were treated so kindly. A member named Joynes who, for fifteen years, had regularly paid the premium on the life of Isaac Cook forgot to pay in the year when Cook died. He pleaded that 'it was peculiarly unfortunate for him that the man should die in the very year it was omitted' and explained that he was bound to pay half the sum insured to various other people. He could forego his own 'moiety' but might the Equitable, he suggested, allow him enough to pay the others. If they did, he wheedled, the public would 'take a favourable idea of an office . . . who have generously not . . . bound an individual to the strict letter of the law'. Joynes was effectively inviting the Equitable to be seen as a soft touch. His request was turned down.[13]

When John Williams of Fetter Lane took his own life the Court refused payment as a matter of principle. In this case one member challenged the decision and a motion to pay was passed. Three months later the motion was quashed. Suicide remained a bar to claiming.[14]

An episode which particularly distressed William was a case of embezzlement in 1815. One of the clerks in the office, James Jones, was responsible for taking the Society's cash to the Bank of England, a temptation too far, and eventually a 'serious deficiency in the accounts of the Society was flagged up by the Bank'.[15] William had not long returned from a visit to friends in Yorkshire when the fraud was discovered and, as he told his cousin, John Price, vexed him very much. He felt some responsibility and possibly he was rebuked by the Court of Directors for not noticing the deficiency sooner. 'I am not so young as I was forty years ago', he wrote to Price. 'I therefore do not [manage] these things so well as I did.'[16]

He admitted in the same letter that he would like to retire but was hanging on so that his son, William, could succeed him. William (junior) had become a clerk at the Equitable in 1809 joining a staff of only six other members: Actuary (his father), Assistant Actuary, Prime Clerk and three existing clerks. When the Assistant Actuary retired in 1817

William, at the age of twenty-six, took over the post. The appointment may smack of nepotism but was more akin to patronage at the time and would not have seemed improper. Besides, young William was an able mathematician and was by this time undertaking many of the calculations for the decennial report.

By 1812 William was sixty-two and feeling his age, not helped by his daily dealings with death – reminders of his own mortality. In 1800 William had begun a twenty-year project analysing the deaths and causes of deaths of members of the Equitable Society, eventually presenting his records in a table.

William's table gives a picture not just of life expectancy but also of the medical knowledge at the beginning of the nineteenth century. The largest group in the chart died from 'Decay (Natural) and Old Age', most of them living into their seventies, some their eighties. The records of the Society reveal more detail: Lady Mary Mordaunt died aged eighty-two in 1819 of 'extreme debility from age', and Martha Hoffman at eighty-seven of 'lethargy'. For some in the 'Old Age' group death was ascribed to 'visitation from God'.

Another large group is those dying from apoplexy. We should now attribute these to strokes, though they would have included all deaths of those in a coma and were probably exacerbated by the amount of alcohol consumed at the time. Dropsy and dropsy-in-the-chest accounted for more than ten per cent of all deaths. Today these would probably be recognised as oedema due to congestive heart failure.

Consumption claimed eight per cent, a number of them young lives probably leaving a family in much need of the annuity from the Equitable. There are several different types of fever, inflammation of various organs of the body, then gout, palsy, quincy (*sic*) in a long list of complaints which would have a more precise diagnosis today. Alongside them are familiar names: cancer, asthma, suicide and, in the Society's records, there is even 'exhaustion by fatigue of business'. Stress is not a modern illness and is one with which William was familiar albeit under a different name.

From an actuarial point of view William's analysis of deaths opens a window on a particular class – the middling sort – and shows, as

TABLE, shewing the *Disorders* of which Persons assured by the Equitable Society have died during the last 20 years, from 1800 to 1821.

Disease.	10 to 20.	20 to 30.	30 to 40.	40 to 50.	50 to 60.	60 to 70.	70 to 80.	80, &c.	Total.
Angina Pectoris	—	—	5	11	12	9	4	3	44
Apoplexy - -	1	3	19	38	69	69	98	5	999
Asthma - -	—	—	—	2	19	19	11	2	53
Atrophy - -	—	—	3	4	6	11	1	—	25
Cancer - -	—	—	1	4	10	8	1	1	25
Child Birth - -	—	—	2	2	—	—	—	—	4
Consumption -	2	9	34	31	44	28	5	—	153
Convulsion Fits -	—	—	3	4	1	3	—	—	11
Decay (Natural) and Old Age	—	—	—	—	5	72	127	58	262
Diabetes - -	—	—	—	2	2	—	1	1	6
Dropsy - -	1	—	7	28	38	41	20	2	137
Dropsy in the Chest	—	1	3	18	34	28	16	—	100
Dysentery - -	—	—	1	1	2	4	4	—	12
Disease of the Stomach and Digestive Organs	—	—	5	4	8	8	1	—	26
Diseased Liver -	—	2	5	24	23	21	4	—	79
Disease of the Bladder and Urinary Passages	—	—	2	4	15	23	15	—	59
Epilepsy - -	—	—	2	3	2	1	2	—	10
Erysipelas - -	—	1	—	2	3	2	2	—	10
Fevers, General -	—	6	18	33	33	39	15	2	146
—— Bilious -	—	1	4	8	9	4	1	1	28

TABLE—*continued.*

Disease.	10 to 20.	20 to 30.	30 to 40.	40 to 50.	50 to 60.	60 to 70.	70 to 80.	80, &c.	Total.
Fevers, Nervous -	—	3	3	13	5	9	3	—	36
—— Inflammatory	—	—	—	4	5	3	2	—	15
—— Putrid	—	2	7	4	5	7	—	—	26
Gout - - -	—	—	1	4	4	11	6	—	26
Inflammation of the Bowels -	1	2	11	13	15	25	9	1	77
Inflammation of the Lungs -	—	—	9	4	24	22	12	2	73
Inflammation of the Brain - -	—	3	7	5	5	3	—	—	23
Inflammation of the Chest, and Peripneumony	1	1	1	1	6	7	4	1	22
Palsy - -	—	1	3	8	26	42	34	2	116
Quincy - -	—	—	—	1	1	1	—	—	3
Rupture of a Blood Vessel - -	—	—	7	14	13	12	3	—	49
Slain in War -	1	1	1	1	—	—	—	—	4
Stone - -	—	—	—	—	1	2	4	1	8
Suicide - -	—	1	2	3	7	2	—	—	15
Water on the Brain	—	—	—	1	3	1	—	—	5
	7	37	166	299	458	536	(345 - - 82) 427		1,930
Number assured during the last 20 years -	1,494	8,996	33,850	45,429	36,489	19,042	6454 from 70, &c.		151,754

THE END.

FIGURES 29A AND 29B William Morgan's Nosological Table, printed as a postscript to *The Principles and Doctrine of Assurances, Annuities on Lives, and Contingent Reversions, stated and explained* (London: Longman, Rees, Orme and Brown, 1821). *(Shown with permission of the Institute and Faculty of Actuaries Library)*

John Wade, a contemporary reformist, pointed out 'the influence of favourable circumstances in prolonging life'.[17] William himself was aware that the mortality experience of policyholders at the Equitable was regarded as being lighter than that suggested by the Northampton Table.[18]

His confidence in the Northampton Table was unshakeable and underpinned his advice to the government when, in 1808, they decided to issue life annuities as a way of raising funds to deal with the debt burden of the Napoleonic wars. The bulk of the Equitable's business at the time was in end-of-life policies; they sold so few annuities that the inaccuracies of the Northampton Table were not immediately obvious. In fact, by assuming too high a mortality, the government's prices for annuities were too low, and therefore much more profitable for the purchaser than the issuer. William's advice cost the government dearly, but he refused to recognise this and was indignant when, in 1819, John Finlaison, a civil servant at the Admiralty (in 1848 to become the first President of the Institute of Actuaries), first forecast the losses that the government would incur and then in 1829 produced an analysis of the situation.[19] There were by this time other mortality tables based on more up-to-date and more detailed data, but William was scathing:

> By the assistance of these tables, the notable discovery has been made of the loss sustained by the public of many thousands every week, for several years past, by granting annuities on lives, computed from the Northampton Table of Observations, which had the surprising effect of alarming the House of commons into a vote for immediately repealing the law which authorized that measure ... With the least knowledge of the subject, it might easily have been seen, that the use of the Northampton Table could not possibly have been attended with such a loss.[20]

With characteristic impatience he gave a caustic swipe at parliament – 'an assembly, supposed to contain the united wisdom and talents of the nation', but 'ill informed' and 'easily misled'. Advancing years had not mellowed his 'Welsh temper'!

FAMILY CELEBRATIONS AND CALAMITIES

As he entered his sixties William's health was uneven. He suffered from bouts of rheumatism and from various unnamed illnesses. Holidays, as well as being family occasions, seem often to have been attempts to restore his health. The benefits of sea-bathing were extolled by the medical profession – with some caveats:

> the advice of a physician . . . should always be taken before a valetu-dinarian commences a course of bathing, either in fresh or salt, hot or cold water . . . Twice or thrice in a week is amply sufficient; and instead of continuing long in the water, or taking repeated dips, the first plunge is the only one that can be attended with any utility.[1]

George III enjoyed trips to Weymouth and went there on the advice of his doctors. A cartoon of 1789 shows him naked and with shaven head (common for lunatics at the time and a reminder of his episodes of madness). A diplomatic ripple in the water protects his modesty as he takes a dip from a bathing machine, whilst a band, thigh deep in the sea, gives a musical accompaniment. The Prince of Wales famously preferred Brighton, where he built his extravagant Pavilion, and there were plenty of other popular seaside resorts. In 1803 *A Guide to All the Watering Places and Sea-bathing Places* appeared, in which (regularly updated) the author, John Feltham, listed the amenities of each place, recommending places to stay and indicating the costs. In fashionable

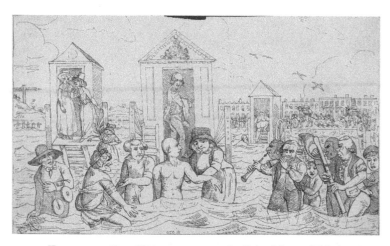

FIGURE 30 *Royal Dipping*, cartoon by John Nixon (1789).
(*BM Satires 7544*)

towns visitors could expect to find libraries (with the latest magazines and journals), assembly rooms, card and tea rooms, banks, a theatre, and a post office with a regular mail service to and from London, as well as shops selling luxury goods. Strict rules about dress and behaviour in public rooms were enforced so that the 'utmost decorum' might be maintained. Outside there were promenade areas, sometimes a bowling green or a cricket ground; there were pleasure boats and, of course, the beaches and bathing machines. These wheeled carts with wooden or canvas walls allowed those bathing to enter in their street clothes and change before the machine was rolled into the water – usually by a horse, sometimes by human power. Bathers emerged from a different door and went down steps into the sea. Women wore voluminous bathing suits; men, until well into the nineteenth century, were allowed to bathe naked, but the sexes were segregated and the emphasis (except for poor George III) was on privacy.

In September 1804 William and Susanna spent a fortnight at Ramsgate, according to the guidebook, a place 'filled with very respectable, and even more select company' than its rival, Margate.[2] Sarah and Susan, their daughters, stayed at Craven Hill with their widowed aunt, Ann Morgan. Despite William's disapproval of Ann's plans to emigrate to America, he seems for once to have held his 'Welsh temper' sufficiently in check for the families to remain on good terms. William and Susanna's sons, William aged thirteen, John, seven, Cadogan, six, and three-year-old Arthur, went to Ramsgate with their parents and the family party was joined by a number of friends, amongst them Samuel Boddington and his children.

William was bored. Ramsgate had less appeal than his beloved Southerndown and Ogmore, their beaches untamed by paths and promenades. For him the two weeks were, he wrote to Sarah and Susan, 'but one dull round of bathing – sauntering – walking on the Pier – looking at the Sea – & talking of the weather; so that when you have the history of one day, you know all that passes here during the season'. In fact the holiday was not without incident. He writes of a three-day gale and a shipwreck on the nearby Goodwin Sands with the loss of ten of the eleven-man crew. He gives a gloomy health report. Susanna, 'as usual when from home, [had] been very poorly', whilst his 'complaint' had returned 'with some violence'. But then he admits that they are both better and bathing every day.[3]

William gives a rare glimpse of Susanna when he tells his daughters that their mother has gone to Dandelion in Samuel Boddington's 'sociable'. A barouche-sociable was the open-top sports car of its day – a four-wheeled carriage drawn by a single horse or a matching pair. It had folding hoods and the occupants could sit facing each other as they bowled along the road. Dandelion, once the site of a twelfth-century manor house owned by the family of Dent de Lyon, was on the outskirts of Margate and had become a public place, with 'shrubs, flowers, a bowling green, a platform for dancing, an orchestra, and other accommodations erected . . . for the entertainment of company who often drink tea in this Elysian spot'.[4] Clearly it was *the* place to go, and Susanna, thanks to Samuel Boddington's largesse, had the added pleasure of arriving in a stylish and fashionable carriage.

For all William's reservations, it doesn't seem such a bad holiday, and seven-year-old John gives his sisters a much more upbeat account of the trip. His letter, exemplary in its careful composition and neat handwriting begins with a slightly priggish tone. He has collected some

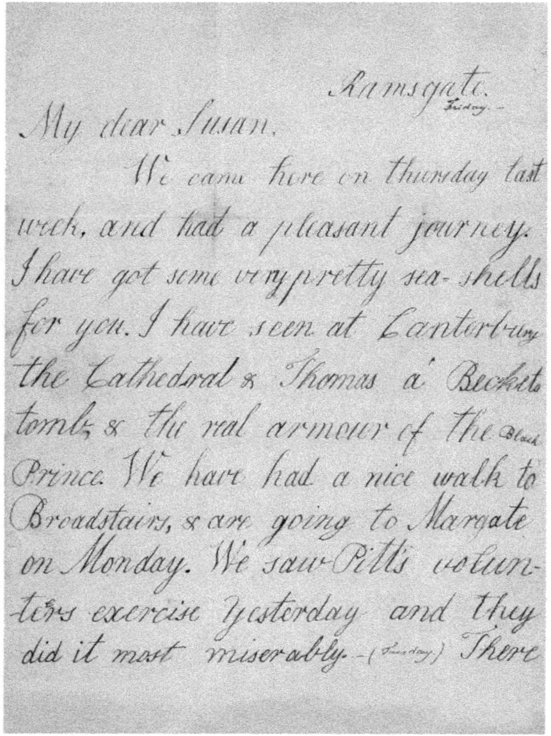

FIGURES 31A AND 31B
Letter of 26 September 1804 from John Morgan to his sister, Susan, written when on holiday in Ramsgate.

John was seven years old at the time (eight the following January).

Ramsgate.

My dear Susan.

We came here on thursday last week, and had a pleasant journey. I have got some very pretty sea-shells for you. I have seen at Canterbury the Cathedral & Thomas a' Becket's tomb, & the real armour of the Black Prince. We have had a nice walk to Broadstairs, & are going to Margate on Monday. We saw Pitt's volunteers exercise yesterday and they did it most miserably. There was a ship lost in the goodwin sands last night. I have not had time to do many lessons, being so much engaged with company, but I hope to make it up when I get home. We see the French coast very plain, but do not care a fig for the French. I have only bathed once in the machine but several times on the shore which I like better. I have got a great deal more to tell you when I come home. My love to Sarah my Aunt & all my cousins.

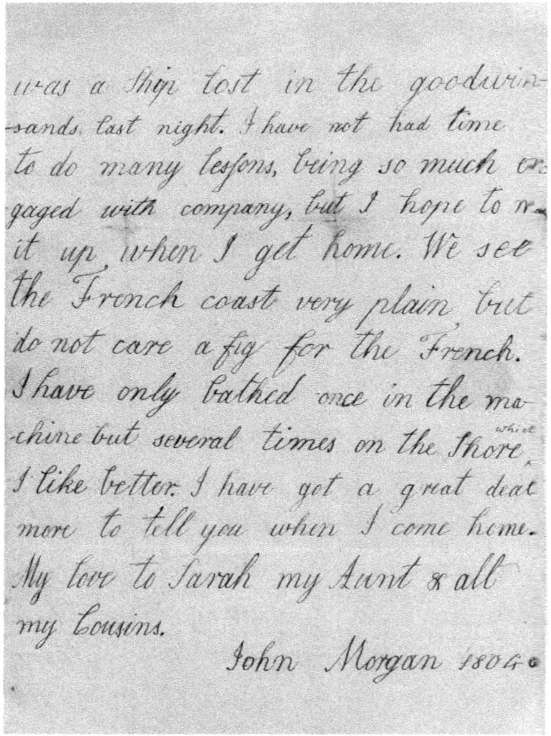

John Morgan 1804

very pretty seashells for Susan but has not had time to do many lessons, as he has been 'so much engaged with company' – he hopes to make it up when he gets home. 'We see the French coast very plain', he writes, 'but do not care a fig for the French' – a received opinion for a child, and a reminder firstly that post-revolutionary France was very different from the idealistic dream so many had welcomed in 1789; and secondly that the 1802 Peace of Amiens had been short-lived. England had been at war with France again since May 1803.

It is more likely his own opinion when he is dismissive about Pitt's volunteers, whom he has seen going through their exercises which they did 'most miserably'. And for all its well-drilled penmanship, the letter does reveal genuine feelings and imagination fired. He has seen at Canterbury 'the Cathedral, Thomas a Beckets [*sic*] tomb and the real armour of the black Prince'. As for the sea, he likes bathing on the shore better than in the machine.[5]

At the end of September the Ramsgate party returned to London, where Sarah Price Morgan and Luke Ashburner were busy visiting relatives and friends before their departure for India. William immediately invited Sarah to dine with him at his office – another sign of cordial relations between the two families.

William was one of the two witnesses to sign the marriage certificate when Sarah and Luke were married at St James's, Paddington, on 15 November 1804, Sarah just a month short of her twentieth birthday. The other witness was Samuel Boddington, who, generous as ever, threw open his house in honour of the newly-weds, where the guests were 'of the fashionable order' and with 'many very intellectual individuals ... prominent among them'.[6] Given his own painful divorce from Luke's sister, it cannot have been easy for him. And did mingling with the lively and stimulating company give Ann Morgan any pause for thought about emigrating to America? Whatever misgivings she may have had, she stuck to her plans. She was to lead a somewhat nomadic life over the next four years, taking her household away from her Morgan in-laws, first to Suffolk and then to Gosport before she sailed in 1808.

Events in family life ran alongside news of warfare, brief and brittle allegiances between European neighbours, and the machinations of domestic politics. On 21 October 1805 the navy, under Nelson's command, defeated the French at Trafalgar. Reports of the Glorious Victory reached London in early November, but with them the news of Nelson's death. The nation mourned; silent crowds lined the streets on 9 January 1806 as his funeral procession went from the Admiralty to St Paul's, where a congregation of 7,000 watched as his coffin was lowered by a winch into the crypt below the dome.

Later in the same month Pitt, exhausted by work and alcohol, died at the age of forty-six. Once again the nation mourned; even his political opponents were generous in their tributes and in their agreement that Pitt's personal debts of over £40,000 (about £2 million in today's money) should be met by a grant from parliament. We have no record of William's feelings about the death of a man he disliked and whom he blamed for many of the country's problems.

Political reform was still a long way off. The Test and Corporation Acts, requiring government officials to take communion in the Church of England, were still in place, as were Pitt's Gagging Acts, and in 1799 and 1800 the Combination Acts were passed. These prohibited trades unions and collective bargaining – in part a response to Jacobin activity and the fear that workers would strike during conflict and force the government to accede to their demands. Reformists could, however, take encouragement from one far-reaching achievement in 1807 – the Act for the Abolition of the Slave Trade.

Perhaps in 1807 the Morgan family was less concerned than usual about politics as they celebrated the marriage of their elder daughter, Sarah, to Benjamin Travers. The union was a most appropriate one, not least in continuing the Morgan family's long connection with the medical profession. Benjamin Travers, already Demonstrator of Anatomy at Guy's Hospital, was at the start of an illustrious career. His CV was impeccable: he had been articled to Sir Astley Cooper who in turn had been a pupil of Henry Cline, the contemporary and fellow student of William in his days at St Thomas's Hospital. Cline was the doctor who attended Horne Tooke when he was imprisoned in the Tower of London and celebrated his acquittal with an annual dinner for friends and supporters. Here was a chain of connections to delight William.

Sarah Morgan and Benjamin Travers were married on 22 July 1807 at St Lawrence Jewry, a handsome Wren building beside the Guildhall and the official church of the City of London Corporation. Their first child, a daughter (another Sarah), was born the following year. William and Susanna rejoiced in the arrival of the granddaughter and, since Sarah had two more confinements in the next three years, the little girl spent much time playing in the gardens at Stamford Hill. But three closely spaced pregnancies took their toll on Sarah's health. Fearful of the future, she begged her sister, Susan, if the worst should happen, to care for her daughter.

Sarah Travers (née Morgan) died on 24 March 1811; she was twenty-six. Grief-stricken, William and Susanna focused their attention on their granddaughter and, in the bleak weeks following her mother's death, they gathered her into their household. Her two baby brothers were too young to grasp the loss of their mother but, for three-year-old Sarah, it was different. At Stamford Hill she was surrounded by familiar faces, in particular her Aunt Susan whom she adored and who in turn adored her little niece. As for the three younger boys, John aged fourteen, Cadogan, eleven, and Arthur only ten, they were not so very different in age from Sarah – more big brothers than uncles and very protective of their new 'sister'.

By July, only four months after her mother's death, she seemed sufficiently settled for Susan to leave her – albeit with some misgivings – whilst she visited friends in Eastbourne and Tonbridge Wells. John, asked to send a report to the anxious aunt, was reassuring. Sarah was busy tending Arthur's cat with milk after breakfast and meat after dinner. A trip was planned so that she might see her brothers, Ben and William, and another to play with a friend. 'She is very happy here without you', wrote John, adding, with complete disregard for punctuation, 'she often asks when you are coming home she will be very much disappointed if you come home without a new Doll for her.'[7]

William and Susanna watched their granddaughter's progress with relief. Being surrounded by her loving aunt and uncles had taken the sting out of her mother's death and she was happy. The sensible course of action, they decided, was for little Sarah Travers to come to live permanently at Stamford Hill under their guardianship.

Benjamin Travers disagreed. His letter to William is lost but it is clear from William's impassioned reply that his son-in-law had demanded that Sarah be moved immediately to his house in New Court, St Swithin's Lane, in the heart of the City. With that 'Welsh temper' roused, William's language is heavy with doom-laden emotion: he is filled with 'grief and astonishment', others at Stamford Hill feel 'the most painful regret and anxiety'. He fears for the health of his little granddaughter and goes as far as to predict that, should she be taken from the open air of Stamford Hill to be 'cooped up in a close room in London ... she will soon follow her poor mother to the grave'.

William saw Benjamin Travers as intending to act precipitously, a move which he warns will 'aggravate the wound which it inflicts with tenfold acuteness'. Not only will Sarah be plunged into misery but the separation will cause distress to her Aunt Susan. Finally, William reminds Benjamin Travers that his late wife had 'assigned her little girl, in the case of her decease, to her sister'.[8]

William won the day; Sarah remained at Stamford Hill.

William had acted in what he believed to be his granddaughter's best interests and Sarah was happy – but at a cost to Benjamin Travers. William had been devastated by the death of his daughter, and yet he expected Travers to part with his infant daughter, effectively putting him through the same bereavement. In fact, a double bereavement because Travers had lost his wife. The bitterness between the two men never healed.

Sarah's father, Benjamin Travers, remarried two years after her mother's death. His second wife, Caroline (née Millett), bore him a daughter the following year and then five more children at two-year intervals. Caroline died in 1829, aged only thirty-nine. And two years later Travers married again – to Mary Poulett Stevens. This marriage ended in separation, but not before three more children had been born. John's letter makes it clear that Sarah had seen her two brothers regularly before the rift between her father and grandfather. Did she continue to see them and did she meet her eight half-siblings? Perhaps not.

There is plenty of evidence that the Travers–Morgan feud continued. For a start, her mother's grave. Sarah Travers (née Morgan) was interred in the Morgan vault at St Mary's, Hornsey. It is a chest tomb with inscriptions on the sides and ends. At one end are the words: 'In memory of SARAH eldest daughter of William and Susanna Morgan and wife of BENJAMIN TRAVERS by whom her remains were removed from this vault and interred at Hendon, Middlesex. She died on 24th March 1811 aged 26.'

The Travers tomb at St Mary's, Hendon, has long since gone; that part of the graveyard was cleared to provide play space for a neighbouring school. But at the London Metropolitan Archive there is a faculty granting permission from the Bishop of London for Travers to re-inter his wife's remains.[9] The date is a shock: 18 May 1829 – just six days after the death of his second wife and a day before her burial. Benjamin Travers buried his two wives side by side on the same day (or within a few days of each other) despite an interval of eighteen years between their deaths. According to the faculty Travers had purchased the vault after his first wife's death, but even so, the double burial is as macabre as it is sad.

The inscription on the Morgan tomb recording the re-interment bears witness to William and Susanna's re-awakened grief at the removal of their daughter's remains. What about Sarah? She was twenty-one in May 1829. Was she present for her mother's re-interment?

It seems unlikely; the family feud continued until Travers's death in 1858. In his will he made careful and detailed provision for all his children – except Sarah, whom he explicitly excludes:

> my said daughter Sarah Travers having from her childhood estranged herself from me and considering her to be already provided for by her Mother's family with whom she has resided from her early infancy it is not my intention to make any provision for her by this my Will.

The kindest interpretation of this decision is that it is practical, but the words are unforgiving, and raise questions. Did anyone at any time make an attempt to end the estrangement?

Sarah's will, some twenty-four years later, included bequests to the children of her two brothers and to some of her half-siblings and their children, showing that she came eventually to know her Travers relations. Inevitably, however, she grew up more Morgan than Travers. Even the house where she lived for the latter part of her life she named 'Bryntirion' after an area of Bridgend, obviously a place which was dear to her. Paul Frame, Richard Price's biographer, has a copy of William's *Memoirs of Richard Price* originally given by Sarah to one of her nieces and described in her inscription as the 'life of one of the best of men'. Her values were those of her grandfather.[10]

Sarah's childhood was happy and comfortable. William was wealthy enough to provide all she could want. She ran free in the garden, kept pets, including a canary and a parrot, and by the time she was sixteen had her own pony. She became the darling of the Morgan family, indulged but, judging by the gentle teasing of her uncles, not spoilt. Her love of all animals was the subject of much family amusement. They are scornful about the parrot, sometimes her 'ungrateful bird' not missing his mistress when she is absent, at other times refusing to speak. Arthur, when she is away, writes to report on the 'various animals with which you have thought proper to burden the House at Stamford Hill and annoy its inhabitants'. John, in one of his letters to her, adopts a mock-heroic style to tell her of the death of one of the pigs: 'cut off in the flower of his youth by a mortal disease – the Husk – nor could all the medical assistance of Dr Scott and myself prevent the vital spark from becoming extinct'.[11]

John's letter is to an address in Brighton where Sarah, Susan and Arthur were staying together with William and Susanna. John tells her that Arthur in his new velvet collar and she in her new spencers[12] must be 'the most dashing couple in Brighton'. She was not more than ten at the time so John's tease is innocent enough, but with only seven years difference in age between Sarah and Arthur it is not surprising that they grew close, and Arthur's letters to his niece seem to show more than avuncular affection.[13] Aged sixteen, Sarah is his 'Ninny Nozzle', and as late as 1831, when she was twenty-three and Arthur thirty, he addresses her with a string of ridiculous titles: she is 'soi disant Princess of Belgium and Antwerp, Radical Professoress of all the learned languages of Europe, dead and living, Professoress of the harp, protectress

Figure 32 Portrait of John Morgan aged thirty-three
(signed but too faded to be accurately read – 'Jack Slader'?)

of tame Animals, Defendress of Thieves and Vagabonds and varmint in general'. The list hints at private jokes and a tender intimacy between them as does the rest of the letter, written in a large flamboyant hand and making extravagant apologies for not attending sooner to her commands. What did William make of the special fondness of his son and granddaughter? Sarah's life is as much his legacy as her story and he had time enough to reflect on the path he had chosen for her.

John Morgan's pencil drawing of Sarah Travers, now framed and hanging in my study, greets me every day but she sits with her eyes cast modestly to one side; she cannot meet my gaze. I have a photograph of her in old age – she lived to eighty-nine but never married. She is sitting outside, a shawl round her shoulders and a lace cap on white hair. The photographer has taken her in profile; she looks serene – relaxed and peaceful – but inscrutable. As for Arthur, he married at the age of thirty-two but he and his wife, Lucy, had no children. A later photograph

FIGURE 33 Undated pencil drawing of
Sarah Travers by her uncle, John Morgan.

FIGURE 34 Sarah Travers in old age.
She died, aged eighty-nine, in 1897.

shows a white-haired Arthur sitting at a desk. There is just a shadow of a smile on an otherwise solemn face but it is hard to reconcile this solid Victorian gentleman with the sparky young man who wrote letters of such jokey fondness.

FIGURE 35 Unattributed photograph of Arthur Morgan FRS.
(every effort has been made to trace the copyright holder and to obtain their permission for the use of this picture)

A COSTLY PEACE

The next greatest misfortune to losing a battle
is to gain such a Victory as this.
(The Duke of Wellington[1])

Politics coloured William's world and impinged on his domestic life, not least with occasions such as his Sunday evening gatherings when he and his friends sang revolutionary songs behind closed shutters, but in the surviving family letters he rarely refers directly to public affairs. Napoleon's abdication and exile to Elba in April 1814, however, was momentous enough for William to write to John Price of his delight in 'the Glorious news and the sure prospect of peace'.[2]

The whole nation rejoiced at the news. The country had been at war for twenty-three years; many, including the three youngest of William's six children, had never known peace; and, for everyone, Napoleon was a bogeyman. His image by this time was mocked not only in cartoons but also on mugs and jugs – bravado in mass production, but he was no less feared for being made a figure of fun.

William's remarks to his cousin suggest that he would have enjoyed the exuberant celebrations which followed the news of Napoleon's abdication, although he might have questioned the expense. A Chinese bridge with a central pagoda was erected for a firework display over the canal which at the time ran through St James's Park. Green Park became home to a magnificent Temple of Concorde; on the Serpentine there was a miniature representation of a sea battle between the English and the Americans (in which the English were victorious); and in Hyde Park thousands gathered to see James Sadler ascend in his silk hot-air balloon.[3]

The national joy was short-lived. After less than a year Napoleon escaped from Elba, returned to France and re-formed his army. His final defeat at the brutal battle of Waterloo cost both sides dearly in the staggering toll of casualties and deaths, as well as the maimed and disabled.

Peace revealed the economic as well as the human price of nearly a quarter of a century of warfare. William's forebodings proved all too accurate. The historian and reformist, John Wade, writing in 1839, put the cost of the war at £1,111,214,731 and recorded that the currency depreciated by twenty per cent.[4] With the war over the government could cease their expenditure on arms, uniforms and all the supplies necessary to support the troops, but this hurt the manufacturers who no longer had a ready purchaser for their goods. Other countries, also recovering from years of war, were in a similar position which hit trade and employment.

Soldiers returning to civilian life exacerbated the situation as they sought work in an overcrowded and changing labour market. Even before the war ended the enclosure acts were altering the agricultural landscape and, at their worst, turning subsistence smallholders into peasant labourers. The new factories and mills with their steam engines were displacing artisan spinners and weavers, and this led to riots and machine breaking, most violent and widespread in 1811 and 1812. The protestors called themselves Luddites after one, Ned Ludd, who, having reportedly broken two stocking frames as early as 1779, achieved mythic status, his name living on in everyday language.

William, travelling north to visit friends in Scarborough in 1815, must have seen some of the changes in the countryside and towns but none of his comments survive. He would have been well aware of the country's economic hardships following peace but he produced no further appeals to the nation. His mission had been to warn; he did not gloat about having made correct predictions. Instead in 1815 he published his *Memoirs of the Life of Richard Price*. Price had intended writing an autobiography but he never got beyond assembling some notes, which have not survived. Then, despite the atmosphere of suspicion and recrimination in the 1790s and the antagonism towards Price, George Morgan began work on a biography of his uncle. If George's other publications are anything to go by, it might have been a very fiery account of Price's life, but George's tragically early death in 1798 left the work unfinished. When William decided to take over the project he found George's papers in a 'confused state' and, more problematic, written in an 'indistinct shorthand'.

'I was reluctantly obliged', William explains in his Preface, 'to give up the investigation [of George's papers] and to take upon myself the task of writing a new, but more concise account.' So the *Memoirs* were in part a work of duty, but above all William's defence of Richard Price's good name and intended, he tells his reader, 'to render justice to the memory of a friend, by bearing my testimony to those virtues and talents ... on which I can never reflect without the deepest gratitude and veneration'.[5]

Respectful, almost reverential in tone, the work is as revealing of William's character and opinions as it is of Price's, to which the references and quotations in this biography bear witness. William frequently uses the first person and it is clear that he shared his uncle's views about the American War of Independence and the French Revolution.[6] William's values are implicit in the praise he lavishes on his subject and in the qualities he admires in him: candour, honesty, humility, modesty, amiableness, benevolence and piety, as well as his ability and industry. Just as revealing are the personal asides he slips into the *Memoir*. There is impatience, even irascibility, in the occasional swipes at people and policies which are of a piece with his famous 'Welsh temper'. The Test Act is 'a reproach to every principle of sound policy and religion';[7] Walpole is 'that great father of corruption';[8] whilst government contracts for supplies for the army and navy are 'shameful bargains'[9] and the result of 'criminal negligence'[10] on the part of the minister responsible. His outbursts are balanced by a generosity of spirit in his mention of Richard Price's friends: Joseph Priestley is 'an admirable philosopher';[11] Benjamin Franklin is one of society's 'brightest ornaments';[12] and the prison reformer, John Howard, is 'amiable and benevolent'.[13]

William limited himself to 'the more important parts of [Price's] life'. In selecting his material he is particularly sparing in his use of private correspondence and scorns the 'indiscriminate publication of letters ... written in the confidence of private friendship'[14] – stern judgement for this biographer to consider!

The following year William published a collection of Price's sermons.[15] In contrast with the energy of his language in his treatment of Price's political views, William is careful and cautious in discussing his uncle's religious opinions. 'It forms no part of my design', he writes, 'to engage in theological disputes.'[16] Evelyn Waugh considered his great-great-grandfather was 'a Unitarian by profession, perhaps an atheist at heart'.[17] It is a plausible supposition. No baptismal records for any of William's children can be found, suggesting that he did not subscribe to conventional rituals. The nearest he gets to stating his views is in a comment in the penultimate paragraph of the *Memoir*:

> On one or two subjects I might possibly venture, though not without considerable diffidence, to express some doubts: but on all the great points in politics and religion, in which I have ever agreed with him, it would indeed be a vain labour to attempt the explanation or defence of opinions which he has enforced with so much truth and energy.[18]

Given his outspoken arguments on other matters, it's a surprisingly mild comment.

Meanwhile the post-war economy zigzagged between boom and bust, and in the acute depression of 1816 even seemed to be mirrored in the weather: ash from the eruption of the Tambora volcano in Indonesia blocked out the sun and the summer, bringing fog, rain and floods. In this gloomy atmosphere members at the Equitable fidgeted and wondered how they might tap into the Society's wealth. The 1810 rule had imposed a five-year waiting period before new members could vote or receive bonuses, but it did not stop some from a restless search for get-rich-quick schemes. One such was a proposal that members of fifteen years' standing should have the option of accepting a transfer of stock in lieu of a cash bonus – an arrangement which would benefit individual members at the Society's expense. A further request for a committee to be appointed to question the Actuary and inspect the books of the Society had an aggressive edge.

The directors asked William for a plan 'to prevent an improper increase of new members'.[19] His suggestion was ingenious: no member should be entitled to vote and no bonus should be given until his or her assurance had become one of the 5,000 oldest policies in the Society, in effect a flexible system linked to the fortunes of the Society. The waiting period lengthened if the number of new assurances grew, shortened if they decreased.[20]

Once again William had put in place measures to secure the future of the Equitable. He had been Actuary at the Society for over forty years – years of growth and prosperity. The General Court wanted to show their appreciation and they did so at a ceremony on 5 June 1817, when Sir John Silvester, Vice-President of the Society, presented him with a silver vase. The payment of £210 to Rundle and Co for the vase indicates that it was quite a substantial size, as does the very lengthy inscription:

> As a gratefull [sic] testimony of the Eminent Services rendered to the Society for Equitable Assurances by William Morgan, Esquire, their Actuary, to the wisdom of whose counsels and to whose indefatigable exertions during more than forty years the unparalleled prosperity of the Institution is in a great degree to be attributed, and to record to his posterity the sense which the Society entertains of those services,
>
> THIS VASE,
>
> was unanimously voted at a General Court of Proprietors on March 6, 1817.[21]

As well as the vase, he was given a gratuity of £1,000 – nearly half his annual salary. William was pleased. He admitted as much in a letter to John Price: 'I have the gratification to know that my labours have not been in vain either to myself or to the Society for which they have been undergone.'[22]

The directors also commissioned Sir Thomas Lawrence to paint William's portrait for the Society (see Figure 1). Lawrence, the leading British portrait artist of the early nineteenth century, painted most of the important personalities of the day in what is described on the National Gallery website as a 'polished and flattering style'. Did he flatter William Morgan? Evelyn Waugh, William's great-great-grandson, comparing the portrait with a profile in ivory showing a large nose and protruding lower lip, claimed that Lawrence 'made the most of his looks'.[23] Perhaps Waugh had a point; a bust of William included in the Royal Academy of Arts exhibition of 1834 confirms his heavy features.

William Morgan was sixty-seven when he sat for Lawrence but his face is surprisingly unlined. The frill of wispy hair above his ears perhaps betrays something old-mannish in him; his breeches, too, old fashioned by 1817 when most men were wearing trousers, are a reminder of his age. The focus of the painting is on the face staring out from a dark background, but the portrait is three-quarter length. Lawrence ends the picture at the knees, tactfully omitting William's club foot. He sits on a gilt armchair upholstered in scarlet with a matching drape above him, badges of success in a portrait commissioned for a boardroom, and yet Lawrence has captured something very human in his subject. He shows him with his left hand held to his throat, the fingers tucked into the collar of his shirt and his thumb resting on his chin. The placing of his hand suggests a very slight tilt to his head and gives him a measure of relaxation. He looks serious yet not without softness and a readiness to smile; Evelyn Waugh felt the portrait lent him 'a meditative, almost poetic air'.[24]

Meditative and poetic? Perhaps, but it is just as likely that William was thinking about the financial advice which Lawrence needed. To the world at large Lawrence – the fashionable portrait painter of the day – appeared hugely successful, but the truth was that he struggled to avoid bankruptcy. To attract the right clientele he needed to maintain a smart studio, but this was expensive. Poor money management and generosity exacerbated his problems and he had to accept loans to keep afloat. Lawrence was probably unaware of the precautions some of his creditors took, but William would have been privy to the fact that the banker Thomas Coutts had in 1802 taken out a policy insuring Lawrence's life for £4,000, presumably seeking to protect a loan he had made to the painter.[25]

What did Lawrence make of his sitter, a shrewd man of business with a blunt manner – very different from the high-ranking military men and politicians he was painting at the time? Was it Lawrence or William who chose to depart from the expected pose of the day?[26]

Eighteenth-century etiquette demanded a stiff, upright deportment and in nearly all Lawrence's many male portraits the stance is conventional of the period; very few have a hand positioned near the face. And was it deliberate that the thumb-to-the-chin arrangement bears a striking resemblance to that of Benjamin Franklin in the portrait which hangs in the White House? Quite possibly William is giving a secret salute. Perhaps a nod to Jerusalem Sols, though by this time defunct, or to the Freemasons, which had survived the purges of the 1799 Unlawful

FIGURE 36 Benjamin Franklin, by David Martin (1767).

The portrait was commissioned by Robert Alexander in thanks to Franklin for settling a property dispute. Franklin is shown reading the deeds belonging to Alexander.
(The White House Collection/White House Historical Association)

Societies Act (and of which Franklin was a member). William's portrait invites speculation that he, too, was a Freemason.

Whether or not it contains a hidden message, the painting celebrates William's achievements at the Equitable. He was successful – and prosperous. His house at Stamford Hill was furnished in fine contemporary furniture; the pieces which have come down through the family are handsome and well made. Census returns show that the Morgans had a butler and other servants. But William did not swank. In a letter to John Price he tells him 'I am sorry you should think it necessary to make an apology for not esquiring me. I like plain Mr Morgan much better.'[27] Nor did he forget his roots. He was proud of being a Welshman and on one occasion turned a Welsh song into elegant English verse on the spur of the moment.[28]

Success meant that the names of the Equitable and of William Morgan spread far and wide – even as far as Scotland, where David Wardlaw, of Gogar Mount in Edinburgh, had the idea of a fund to provide annuities for widows, especially those of the clergy of the Church of Scotland. Early in 1813 David Wardlaw, together with two others, travelled to London expressly to consult William about the way to proceed. William was generous with his time and expertise: he revised the formulae for their calculations and he amended the articles of their proposed constitution. Thanks to William the foundations were laid and the new society was soon up and running. And is still running. Its name is Scottish Widows.[29]

William's knowledge and experience were more formally tapped when he was called upon to give evidence to the 1817 Select Committee on the Poor Law. He answered a raft of questions about the successes and failures of friendly societies, asserting that they had flourished wherever the plan he had recommended had been adopted. He confirmed the accuracy of Price's tables for calculating the contributions necessary to allow annuities for all contributors and he discussed the insurance risks attached to some occupations. 'We object to some trades', he told the committee, 'particularly to painters, publicans and bakers; the latter because they sleep over their oven.' He added that they objected 'rather to plumbers' but gave no explanation for this.[30]

Neither the weather nor the economy improved in 1817. Another wet, cold summer; another sluggish year in the money market. On the domestic front William continued to attend to his cousin's financial affairs but, by August 1817, John Price's health was failing and the letters are addressed to Price's wife, Jane. His advice is tinged with ill-concealed tetchiness. He tells Jane that he has, as she wished, sold an Exchequer Bill purchased only three months earlier. With the £500 realised by the sale he bought two others for £200 each, leaving a balance of £106 7s. 3d. He transcribes the broker's account so that she can see 'how very unprofitable this method of laying out money is' and rams home his point: her money 'might as well have been locked up in [her] closet'.[31] It is as much a comment on the national economy as on John and Jane Price's personal finances. The slump continued. A few months later he writes: '[Government] funds are at present in a state which makes it highly improvident (in my opinion) to invest any money in them.'[32]

The stagnant economy exacerbated unemployment which led to unrest and even bloodshed in the cause of political reform. In 1819 a mass meeting at St Peter's Field in Manchester was violently broken up by the cavalry, resulting in eleven deaths and at least four hundred injured: the 'Peterloo Massacre' – this headline in the *Manchester Observer* of 28 August 1819 labelled the outrage and, with its sardonic reminder of the battle of Waterloo, signalled war between the disenfranchised and an oppressive government. The newspaper's report claimed to be 'a full, true and faithful account of the inhuman murders, woundings and other monstrous cruelties exercised by a set of INFERNALS (miscalled Soldiers) upon unarmed and distressed People'.

The government tightened its grip by passing the 'Six Acts' which, as well as prohibiting meetings of more than fifty people and imposing heavy penalties for seditious libel, increased the stamp duty on newspapers and cheap pamphlets to 4d – a direct hurt to radical publications. Reformists were muzzled.

William made no public comments about the political situation. His letters to John and Jane Price might have included private remarks, but by 1819 both had died. The immediate post-war years were momentous for William personally: in 1818 the marriage of his eldest son out-shone events on the national stage – a marriage which could not have been more fitting. Beautiful Maria Towgood, William's bride, was the daughter of Samuel Rogers's younger sister, Martha, and so the union forged another link between old friends. Both families could celebrate again the following year with the birth of a daughter, Frances Maria.

With his son settled in marriage and thriving in his post at the Equitable, William, who was nearing seventy, was ready to hand over the role of Actuary to a worthy heir. It was not to be. In a ghastly echo of his sister's death only eight years previously, William Morgan junior died on 5 October 1819. He was twenty-eight. The cause of his death was 'inflammation of the chest'. If, as seems likely, this was tubercular in origin, his father – medically trained in his youth – might have been expected to recognise the signs of a fatal illness, and his knowledge that there was no cure must have been heartbreaking. So devastated

FIGURE 37 Pencil portrait of William Morgan Jr (1791–1819).
Inscribed on the back and initialled by Sarah Travers:
'This portrait was drawn from memory a few days after his decease by his brother, John Morgan, an excellent likeness'.

was William that he included a reference to his bereavement in the address to the Society just two months later, in December 1819. His own old age, he told the Court, had 'in a great measure been repaired to [him] by the accession of a near and dear relation to participate in [his] labours'. But, he continued, in a bald, bleak statement, 'all my hopes with regard to him are for ever gone'.[33] It's a direct reference to his son and yet it seems he cannot bring himself to speak his name, or even to use the word 'son'.

The untimely death was mourned at the Equitable, where William junior was a popular colleague and able at actuarial calculations. Such was the respect for his work for the Society that the General Court doubled the gratuity for his dependants, raising it from £500 to £1,000.[34] His wife, Maria, widowed at only twenty-six, devoted her life to her child – and to the memory of her husband, for she refused to consider any subsequent offer of marriage.[35] She had another bereavement to face when she was seventy, with the death of her daughter, Frances – unmarried – at the age of only forty-five. Maria survived another fourteen years, dying in 1878, aged eighty-four.

Meanwhile William, though he could not relieve the ache in his heart, had to find a successor at the Equitable; he needed a confidante, someone to support him. His second son, John, twenty-three in the January of 1820, had decided on medicine and already obtained the diploma of the College of Surgeons. Cadogan, aged twenty-one, was set on a legal career. He was in due course admitted to the Middle Temple in November 1823 and called to the Bar on 28 November 1828.

That left eighteen-year-old Arthur – the merry, fun-loving youngest son of the family. From a distance of more than two hundred years he does not seem an obvious candidate; nevertheless in 1820, a year after his brother's death, Arthur joined the Equitable. However much his father may have wanted this outcome, it does not seem likely that he would have exacted obedience to his wishes. William, at the same age as Arthur, had bowed to paternal pressure and agreed to study medicine. His memories of those difficult years, together with his enlightened views on education, would surely make him reluctant to force a career path on his son.

Today we might label Arthur's appointment, like his brother's before, as nepotism. William is more likely to have regarded the appointment of a family member to succeed him as the responsible course of action. Arthur might have seen it as filial duty; he might even have been tempted by an attractive salary, for he began work on the same pay as his late brother had earned at the start of his career. Whatever his motivation, he was to make a success of his choice.

RISE, PROGRESS, MISREPRESENTATION

Surely Mr Babbage must think very meanly of the abilities or
the integrity of those who conduct the affairs of the Society.
(William Morgan[1])

Arthur arrived at the Equitable at a time when the national economy
was still only bumping along. The gold standard was restored in
three stages during the years 1819 to 1821, initially causing deflation
which, in turn, meant high prices and more unemployment, but eventually
leading to recovery and, by 1823, a frenzied boom and wild speculation.

The Equitable's funds, invested in government stock, rose in value
and, predictably, policyholders wanted to cash in on their perceived
windfall. The get-rich-quick brigade became increasingly vociferous,
arguing that bonuses should be added to the sum assured more often
than every ten years, and even that part of the bonus, instead of being
payable at death together with the sum assured, should be paid immediately. These opportunists failed to consider the long term. Money
could certainly be realised by selling investments but, once the profit
was creamed off, the Society's funds needed to be re-invested. Fresh
stock was available, but only at less advantageous interest rates. Similarly,
premiums from new members, vital for the continuity of the Society,
also had to be invested at the current low rates of interest. Risks and
benefits needed to be shared to maintain success.

Never had William Morgan been more in need of moral support.
Arthur rose to the occasion so well that in March 1824 William wrote

to the President, Sir Charles Morgan,[2] recommending promotion for Arthur. His tactful modesty and his pride in his son reveal as much about William as they do about Arthur. He acknowledges that 'the partiality of a father renders him a very unfit judge of the merits of his son', but 'in justice to [his] son ... and perhaps to the Society' he feels he should not be 'altogether silent'. He can, he tells Sir Charles, vouch for Arthur's integrity and his knowledge of the affairs of the Society, and is confident that Arthur 'will never be wanting in diligence or zeal to promote its interests'.[3]

Arthur was appointed Joint Actuary. He was twenty-three, the same age as William had been when he was appointed Assistant Actuary some fifty years previously. Arthur had the benefit of the methodology which William had developed but he took office at a tempestuous time. The years 1824 and 1825 saw the birth of a number of new assurance societies. Many were destined to fail, but in the heady atmosphere of the time some members of the Equitable were selling their policies in public sales, some realising twice their office value. Demands for bigger and more frequent bonuses became more strident.

It is a measure of William's confidence in his son that in the August of 1824, despite the fevered atmosphere, he left Arthur in charge at the Equitable and went to the seaside – to Worthing – together with wife, his daughter and his granddaughter, Sarah Travers. Arthur's letter to Sarah shows him entirely unintimidated by his new position, almost flippant, using as his writing paper the flyleaf of a document for his father. He writes in his usual cheery style, teasing Sarah for her failure to write to him and sending a message to his father that he is 'quite mistaken in supposing that [his] late indisposition arose from taking ale beer fruit pye greens – cream and the like'. He and Cadogan have, he claims, been living like hermits since the family departed.[4]

Arthur mentions in the letter the possibility of his father 'taking a journey to Wales in the Autumn'. Assuming William did make his annual trip to Southerndown in September, he returned to a series of angry meetings at the Equitable. Members clamoured to propose alterations to the existing allocation of bonuses, and meetings became ever more assiduously attended – so much so that the *Times* began to print full accounts of the proceedings. The quarterly meeting on the 2 December 1824, it was reported, was 'crowded to excess' and, judging by the frequent inclusion of 'Hear! Hear!', was restive and noisy.

The principal activists argued for changes which on the face of it were suggestions for more up-to-date governance, but which, if implemented, would have completely disrupted the fabric of the Society; some of their proposed schemes – designed to benefit members of long-standing – would have resulted in putting an end to the Society. William and the directors, as well as recognising the dangers to the Society, abhorred the greed lying behind the proposals. The Society had been founded, the directors stated in their report in March 1825,

> to provide for the comfort of others, to extend blessings and protection to the widow and orphan, and to assist, as far as pecuniary relief would afford it, to lighten affliction and relieve distress, and to procure support for those who have lost their natural guardians and protectors.

It was never the intention, they continued, 'to make it a subject of speculation'.[5]

William, in his address, was equally punchy, if a little weary that once again he had to warn about 'the too sanguine hopes of some of [the Society's] Members' and the dangers of eroding its capital funds. 'On the inviolability of that capital', he declared,

> depend the good faith, the honour and ultimately the existence of the Society. To my former Addresses therefore I must now beg leave to refer the General Court, being well aware that if they shall have lost their effect, I can have but little hope of succeeding by any new arguments, were I possessed of health and spirits to advance them.[6]

The various proposals were defeated and the names of those who criticised the Society have long since faded into obscurity.[7] Except one.

Quite the most painful for William were the comments by Charles Babbage. Babbage, aged thirty-three, had yet to be appointed Lucasian Professor of Mathematics at Cambridge, but he was recognised as a leading mathematician of his generation and he had already started work on his calculating engine. In 1824 he had been invited to become actuary of a new life assurance company, the Protector. In the event, the Protector never got off the ground so Babbage's job did not materialise, but his interest in life assurance seems anyway to have begun at least five years earlier with a 'plan' which he sent to William, requesting his opinion. William refused:

> Equitable As^ce^ Office
> 22nd Octo^er^ 1819
>
> Sir
>
> It would give me great pleasure to attend to the enclosed plan, which you did me the honor [*sic*] to send me; but my office and other engagements have obliged me to decline all business of this kind, and therefore I have returned your paper; knowing that no person can give a more correct opinion upon it than yourself.
>
> I am, Sir,
> > Your obedient servant
> > Will Morgan

The brief letter is a curt rebuff.[8] The date, however, is telling; 22 October was less than three weeks after the death of William's beloved son.

In happier times there might have been friendly dialogue between the two men, who had more in common than mathematics. Charles Babbage, like William, was a radical thinker and ahead of his time in his views on politics and education, not least in watching closely events in France following the Revolution. In particular Babbage admired the establishment of the Grand Écoles, with their emphasis on science, and he approved of the technical high schools of continental Europe, where the value of practical skills was recognised and engineers esteemed. He urged reform of the British education system, arguing that a better understanding of scientific and technical methods would in due course benefit British commerce and industry.[9]

Babbage deplored the fact that scientific knowledge scarcely existed amongst the higher classes of British society. In *A Word to the Wise* he went further, advocating the abolition of hereditary peerage and the introduction of life peers.[10] There are clear echoes of Tom Paine here and William would surely have applauded Babbage's views and arguments.

In actuarial matters things were different. In 1826 Babbage published *A Comparative View of the Various Institutions for the Assurance of Lives*. It is presented as a layman's guide and in part gives a comparison of the numerous assurance societies which had been established in the early years of the nineteenth century. In the preface the Equitable is singled out for praise and Babbage explicitly agrees with 'the sentiments expressed by Mr Morgan'. He quotes William's words: 'I consider every assurance made for the purpose of providing for a surviving family, in whatever office it is effected, not only as a private but as a public good.'[11]

The tone changes in subsequent chapters and Babbage levels sharp criticism at the Equitable's premiums and bonuses, both touchpapers for that 'Welsh temper'. William wrote a detailed letter, first complaining that he could have provided relevant information had Babbage requested it. (He seems to have forgotten Babbage's earlier approach.) In the rest of the letter he deals point by point with the 'several mistakes' which William claims Babbage has made.[12] A year later he published *A View of the Rise and Progress of the Equitable Society*.[13] It is in part a history of the society, but also a full response to Babbage's publication.

Just as Finlaison had done (see Chapter 19), Babbage questioned the accuracy of the Northampton Tables but, whereas Finlaison had been considering the premiums for annuity policies, Babbage focused on those for end-of-life policies. He pointed out that the mortality rate among those with life assurance was lower than the general mortality rate. This, he claimed, coupled with good rates of interest on the money accrued from premiums, meant that the Equitable's premiums were unfair to its members. William had already acknowledged that those taking out life assurance tended to come from a class sufficiently cushioned by wealth and privilege to enjoy better health and prospects than the general population but, in response to Babbage, he argued that the difference between those lives measured by the 'Equitable experience' and the 'general mass of mankind' was significant only in the early years.[14] After reaching forty (the age when most policies were taken out) the difference became less marked.

As for the bonuses, William detailed the mistakes which he claimed Babbage had made, and was outraged at Babbage's accusations that the Society had acted unfairly in keeping back some of the surplus and had misled the public about the proportion of surplus which had been distributed. 'Surely', cried William, 'Mr Babbage must think very meanly of the abilities or the integrity of those who conduct the affairs of the Society.'[15] Babbage had, William claimed, confused the amount apportioned to members with its value. He disagreed with the tables to calculate bonuses which Babbage had produced, and rejected Babbage's suggestion of an annual distribution of bonuses – a course which would keep the Society 'in a perpetual state of poverty'.[16]

There are echoes of the spat between William and Adair Crawford nearly fifty years earlier, but then the biffs and thwacks were veiled in anonymity. This time the exchanges were very open, very public and, for William, very wounding. He was no longer a young hound on the attack; instead, he was an old man defending himself against an energetic and knowledgeable critic, and one with loyal friends. The publication date of *The Rise and Progress of the Equitable Society* was unfortunate in that it was the very year in which the deaths occurred of Babbage's father, his wife, his eleven-year-old son and a newborn infant. Babbage, grief-stricken and close to a complete breakdown, was persuaded to make a tour of the continent as an aid to his recovery. In his absence and on his behalf others took up the fight against William.

On 26 June 1826 the *Times* published a letter from Francis Baily, a long-standing member of the Equitable. His attack was detailed and savage – and well supported with precise references to William's book. William, he claimed, had passed 'unmerited censures' on Babbage and made 'unfounded remarks', with opprobrious constructions and singular perversions of Babbage's meaning. William had insinuated, had misrepresented and treated Babbage unfairly.

William's response, published on 1 July, was uncompromising. With weary patience he argued that Baily had 'put a construction on the remarks [he] made which [he] never intended', and he sought to justify himself by selecting two points in Baily's criticism for detailed examination. His arguments, however, are tenuous and his tone is indignant; his letter seems unlikely to have swayed the existing opinions of the readers.

Unfortunately for William, on the same day that his letter was published another appeared sent by George Farren, a solicitor and founder of the Economic Life Assurance Society. Farren took exception to a perceived allusion to his company and – unsurprisingly – to William's use of 'quackery' in describing the company's method of calculating risk.[17] Farren's letter was long and detailed, and finished by describing William as 'declining into the sear [*sic*] and yellow leaf'; William did not reply or, if he did, his letter was not printed.

The weather did nothing to cheer William as a wet July was followed by storms in August. And criticisms continued. Henry Brooke FRS reached conclusions 'entirely at variance with the opinions entertained by Mr Morgan'.[18] Thomas Young, a well-known and well–respected scientist, wrote a letter, published in the *Philosophical Magazine* for 1828, in which he questioned William's figures – another public humiliation.

A FRAGMENT

For William the dispute seems to have ended on a sour note. Or did it? On an unremarkable beige page of the autograph album (part of the original tea caddy legacy) is Charles Babbage's signature, cut from a letter.

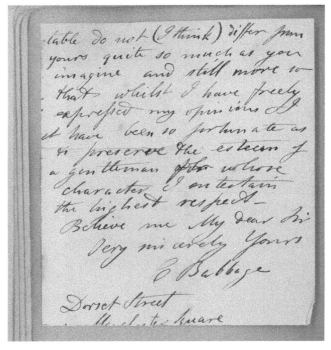

FIGURE 38 Fragment of a letter signed by Charles Babbage, found in an autograph album, part of the author's tea caddy legacy.

It provides an astonishing clue: a scrap of the letter remains and I read:

> [-]table do not (I think) differ from
> yours quite so much as you
> imagine and still more so
> that whilst I have freely
> expressed my opinions I
> [?] have been so fortunate as
> to preserve the esteem of
> a gentleman for whose
> character I entertain
> the highest respect.
> Believe me My dear Sir
> Very sincerely yours
> C Babbage

Equitable is not the only word ending '-table' but the context makes it the most likely and, whatever the first word of the snippet might be, the sense and tone are clearly conciliatory. It raises questions. What was William's reaction? Did he reply and, if so, in what spirit? He would have found it very hard to forgive Babbage. The letter is private but Babbage's book reached a wide audience and his criticism was damaging. Then at what stage in the dispute did Babbage write his letter? Given his depression and his trip to the continent he probably wrote to William early in the debate, before the letters to the *Times* intensified the attack. Nevertheless the surviving fragment shows at the very least that William kept Babbage's letter – some indication, perhaps, of the importance he attached to it.

FINAL YEARS

Morgan ... combined the gift of theoretical investigation
with that of business acumen, a combination which
actuaries in this country regard as the ideal.
(W. Palin Elderton[1])

Babbage's book and the correspondence in the *Times* reinforced the fact that the name of Morgan was, for the public, synonymous with insurance; it even featured in a satirical rhyme which was published in *John Bull*, a Tory weekly paper. The editor, Theodore Hook, used the friendship of William and Samuel Rogers to poke fun at Rogers's very pale and cadaverous appearance as well as his poetry. (Even the title, borrowed from Rogers's best-known poem, is part of the joke.[2])

Human Life

Cries Sam, 'All Human Life is frail,
E'en mine may not endure;
Then, lest it suddenly should fail,
I'll hasten to insure.'

At Morgan's office Sam arrived,
Reckoning without his host;
'Avaunt!' the frightened Morgan cried,
'I can't insure a ghost.'

'Zounds! 'tis my poem, not my face;
Here, list while I recite it.'

> Said Morgan, 'Seek some other place,
> I cannot underwrite it.'

The verses appeared in the issue of 7 March 1824, by which time William was struggling with failing health as well as troublesome members. As early as 1823 he had referred, in a letter to the President of the Society, to a 'very serious and dangerous illness' and, by 1829, he was too frail to attend the General Court. His address was read by the Vice President. Effectively a valediction, the speech recorded William's satisfaction with the measures in place to safeguard the Equitable's funds. After a characteristic swipe at those with 'unreasonable expectations', William made it clear in his concluding sentence that he intended to retire: 'I should be very sorry to take my leave of the Society with any other impression on my mind than that of a full assurance of its continued welfare and prosperity.'[3]

The President, Sir Charles Morgan, wanted William to continue as Actuary as long as he lived, but William asked to be released from his post. His retirement set the directors a problem. How could they reward his fifty-six years of indefatigable service and his pioneering achievement in building a successful life assurance society? They recorded that they regarded his industry and ability 'as the principal cause of the prosperity of an Institution, which has afforded and will continue to afford blessings to thousands'. And they granted him a pension of his full salary of £2,000. They were well placed to know that William's life expectancy was limited, but the award was a generous compliment.

William was gratified. His resignation, he told his youngest sister, Sally, had been 'received with every mark of respect and affection'. He was pleased, too, that Arthur had been elected to succeed him as Actuary and that it had been done 'in the handsomest manner'.[4]

William managed, despite crippling rheumatism and 'with no inconsiderable degree of exertion', to attend the General Court to make his formal farewell. He wanted, he said, to put on record his debt to those who had originally appointed him, and he looked back with pride at the Equitable's rise from 'a puny institution' to a 'magnificent Establishment'. He would 'never cease to be anxious for [the Equitable's] welfare', and he hoped that the Society might 'long continue to enjoy and deserve the approbation of mankind'.[5]

William's retirement at the end of 1830 coincided with an end of era in the nation. George IV had died in June of the same year. A lifetime of over-indulgence had made him obese, dropsical and gout-ridden. Nearly blind with cataracts, he rarely made public appearances in his final years and his death was not mourned. 'There never was an individual less regretted by his fellow creatures than this deceased king', declared the *Times* on 15 July 1830.[6] He was succeeded by his younger brother, William, still sufficiently hands-on to appoint the prime minister of his choice but less inclined than George to meddle in politics.

Throughout the country agitation for a fairer electoral system was growing. Some reforms were already in place. The detested Test and Corporation Acts had been repealed in 1828, giving religious liberty to Dissenters, and the Catholic Emancipation Act was passed in 1829, allowing Roman Catholics to sit in parliament.

William followed events closely. 'I suppose you are gratified by the politics of the day. The borough-mongers are in great dismay at the dissolution of parliament which will take place in a day or two', he wrote to his sister on 21 April 1831. The date is as significant as the comment. A reform bill, having narrowly passed through its first and second readings in the Commons, was defeated in committee on 20 April. 'The Crisis is arrived!' announced the *Times* in its leader on 21 April. Whilst the *Times* as well as William and his sister might have been disappointed at the bill's defeat, they would have been pleased at the speed of the election which followed the dissolution of parliament. In the pocket boroughs where wealthy patrons used borough-mongers to bribe voters, the snap election gave too little time for the usual corruption.

Reform was in the air and eventually a bill was passed in 1832. Parliamentary seats were redistributed so as to reflect the number of voters returning a member; pocket boroughs were abolished; similarly rotten boroughs, those with no voters but which returned two MPs chosen by the landowner. It did not achieve all that William would have wished but it was a start – the first of the three landmark reform

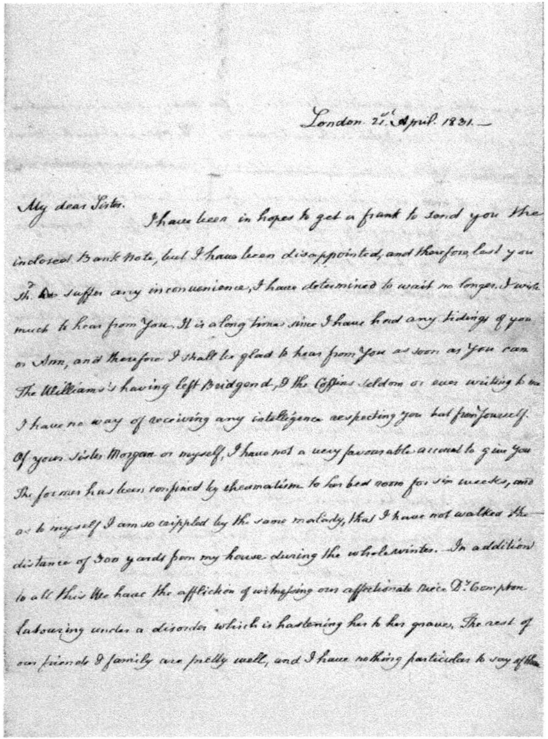

London 21st April 1831.—

My dear Sister.

I have been in hopes to get a frank to send you the inclosed Bank Note, but I have been disappointed, and therefore lest you sh[oul]d suffer any inconvenience, I have determined to wait no longer. I wish much to hear from you. It is a long time since I have had any tidings of you or Ann, and therefore I shall be glad to hear from you as soon as you can. The Williams's having left Bridgend, & the Coffins seldom or ever writing to me I have no way of receiving any intelligence respecting you but from yourself. Of your Sister Morgan or myself, I have not a very favourable account to give you. The former has been confined by rheumatism to her bed room for six weeks, and as to myself I am so crippled by the same malady, that I have not walked the distance of 300 yards from my house during the whole winter. In addition to all this We have the affliction of witnessing our affectionate Niece Dr Compton labouring under a disorder which is hastening her to her grave. The rest of our friends & family are pretty well, and I have nothing particular to say of them.

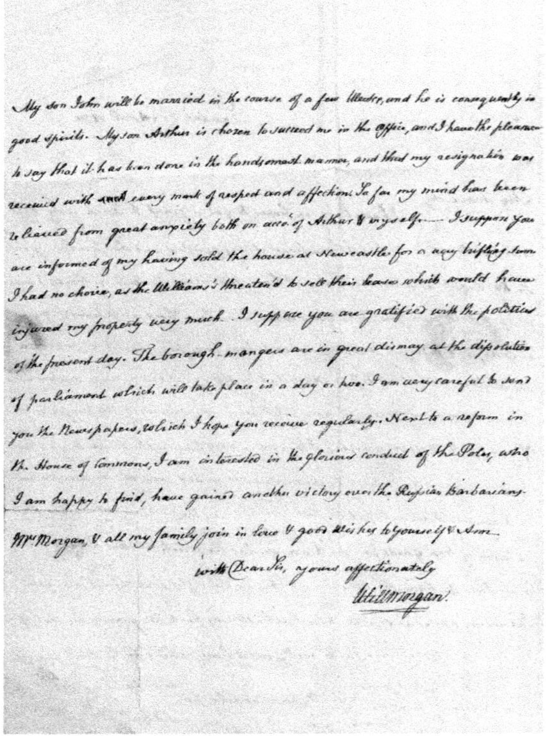

My son John will be married in the course of a few Weeks, and he is consequently in good spirits. My son Arthur is chosen to succeed me in the Office, and I have the pleasure to say that it has been done in the handsomest manner, and that my resignation was received with every mark of respect and affection. So far my mind has been relieved from great anxiety both on acc[oun]t of Arthur & myself.— I suppose you are informed of my having sold the house at Newcastle for a very trifling sum. I had no choice, as the Williams's threaten'd to sell their lease which would have injured my property very much. I suppose you are gratified with the politics of the present day. The borough-mongers are in great dismay at the dissolution of parliament which will take place in a day or two. I am very careful to send you the Newspapers, which I hope you receive regularly. Next to a reform in the House of Commons, I am interested in the glorious conduct of the Poles, who I am happy to find, have gained another victory over the Russian Barbarians.

Mrs Morgan, & all my family join in love & good wishes to yourself & Ann

with Dear Sis, yours affectionately

W. Morgan.

FIGURES 39A AND 39B Letter of April 1831 written by William Morgan to his youngest sister, Sarah Huddy (1761–1831), widow of David Huddy.

Sarah died in May 1831. William outlived all his seven siblings and died in June 1833.

acts of the nineteenth century, though William would not live to see those of 1867 and 1884.

William's main reason for writing to his sister was to send her a banknote (the amount is not specified). He was presumably still dealing with money matters for members of the family and the copy in my collection is probably just that – a copy – showing him to be as meticulous as ever in his record-keeping. It is the last of his surviving letters. His handwriting is as firm and legible as ever, while his remarks show his sharp mind and his continuing interest in international as well as domestic affairs. 'Next to reform of the House of Commons', he writes, 'I am interested in the glorious conduct of the Poles, who I am happy to find, have gained another victory over the Russian Barbarians', a reference to the November uprising of 1830–1 when young Polish officers led a rebellion against the Russian Empire. William's delight in their early success would have been short-lived; the rebels were eventually crushed and Poland became an integral part of Russia.

Did William also keep abreast of scientific affairs? The year 1830 was a critical date, with the publication of Lyell's *Principles of Geology*, in which he claimed that the Earth was very much more than 6,000 years old. His arguments were based on the evidence of geology. Lyell recognised that mountains and valleys, plains and rivers, and the fossils embedded in our world told the story of a slow accumulation of changes over a vast span of time rather than a divine operation lasting only six days.

Lyell's book inspired and devastated in equal measure. For Charles Darwin, who was given volume I when he set out in the *Beagle* in December 1831, it was 'a grand work which future historians will recognise as having produced a revolution in science'.[7] But for others it was a challenge to the authority of the biblical creation stories. 'Are God and Nature then at strife?'[8] cried Tennyson, agonising that 'scarped cliff and quarried stone' as well as 'Nature, red in tooth and claw',[9] supported Lyell's thesis.

William had a direct connection with the debate that followed the publication of *Principles of Geology*. In William's letter to his sister he tells her that John 'will be married in the course of a few weeks, and he is consequently in good spirits'. John's bride was Anne Gosse, a first cousin of the naturalist Philip Henry Gosse, whose name was to become indelibly linked with the clash between theology and science. Religion was as important to Philip Henry Gosse as his scientific studies. Arguably the same could said of Richard Price, the uncle whom William so much admired, but in Price religion was coupled with rational dissent. Gosse expressed his faith with cloying piety. He lived in daily preparation for Christ's Second Coming and in eager anticipation of 'the Glory that is to be revealed'.[10]

Lyell's book, with its contradiction of the stories in Genesis, shocked many scientists but especially Gosse. Worse, his own geological studies confirmed Lyell's work but he could not bring himself to jettison the biblical version of creation. Gosse was in turmoil. It was not just that the Earth was many thousands of years older than had been thought but the fossils carried an even more alarming message: some species had become extinct. What was to stop man from suffering the same fate? Where was God in all this?

Eventually Gosse reconciled his religious beliefs and his scientific observations with an ingenious, if ludicrous, solution. On the day God created Adam, he gave the Earth instant antiquity. As he carved out the mountains and valleys, he buried fossils in the rocks. Gosse published his theory in 1857 in *Omphalos*, the Greek word for 'navel' and an allusion to the belief that Adam as the first man would have had no navel. The book provoked anger and ridicule amongst geologists and theologians; perhaps it is just as well that William had died before its publication.

At his wedding on 6 May 1831 at St James's Church, Poole, John was supported by his brother, Cadogan, but William, aged eighty-one, was too frail to attend the ceremony. Had he done so he would have met several members of the Gosse clan; Anne's parents and siblings are all listed as witnesses. Philip Henry Gosse was in Newfoundland at the time otherwise he might well have joined the party. Only a year older than his cousin, he had seen much of Anne as they both grew up in Poole, and over the years they remained close. There is no evidence that Anne shared Gosse's narrow beliefs but she must have been fully aware of them. She nursed Gosse's first wife, Emily, in her final illness and looked after seven-year-old Edmund for a few weeks after Emily's death.

Edmund Gosse famously depicted his father as austere and joyless in *Father and Son*. It is a portrayal which has been challenged,[11] but Edmund's book, never out of print since its publication in 1907, continues to shape perceptions about Gosse as does *Omphalos*, not forgotten though long since out of print. Gosse's many books about zoology with their detailed illustrations help to balance the picture. In Gosse's careful observation of the natural world, William might well have found some common ground.

William died on 4 May 1833, less than a month before his eighty-third birthday. His son, John, writing to a friend, said that his father had 'fallen victim to the prevailing influenza' but he had died 'without pain or any other suffering'.[12] His death was one of twelve announced in the *Times* on Monday 6 May. Although not all the ages are specified, the list provides a postscript on life expectancy in 1833. Only one, a woman in her eighty-ninth year, was older than William. There were two infant deaths, one a death in childbirth; two were in their seventies, one sixty-one, another fifty-eight.

William had outlived all his siblings. Nancy, the younger sister who had kept house for him at the start of his career, died in 1822. She was seventy. Kitty, the oldest of the family, died the following year at the age of seventy-seven. And Sally, William's youngest sister, died in May 1831, aged sixty-nine. Sally and Kitty were buried at St Mary's, Coity (near Bridgend), Nancy at Llandaff.

William's funeral and burial at Hornsey was, as he wished, quiet and private – unlike his uncle's forty years previously. A brief obituary appeared in the *Morning Chronicle*. In the formal language of the period it praised 'his mathematical and scientific attainments [which] were of the highest order' and noted his fifty-six years at the Equitable and its rise to become 'an establishment of national importance . . . diffusing its benefits to thousands of families, and securing them in the enjoyment of comforts of which they would otherwise have been rendered destitute by the death of their friends and relations'. The tribute included a reference to William's outspoken political views:

On the subject of public credit and the national debt he was one of the most popular writers of his time, never hesitating, in his public writings or in private conversation, to state his opinions on those subjects with the utmost freedom, and to express his unqualified disapprobation of the financial administration of Government in regard to the terms on which loans for the public service were negotiated and contracted for during the whole period of the late war. He was the steady and ardent friend of civil and religious liberty, and nearly the last of that band of philosophers and patriots who lived during the American and French revolutions, the associate of Franklin, Price, Priestley, Horne Tooke and many other eminent men.

It is a neat and not ungenerous assessment of the public man but it does not touch on his modesty, his kindness — particularly in giving advice to fledgling assurance companies – and his amazing stamina in the innumerable complex calculations he undertook. Nor does it deal with the private man, hospitable and warm-hearted, and a proud and loving father.

AFTERWARDS

In this vault beneath are deposited the remains
of WILLIAM MORGAN of Stamford Hill.
He was born at Bridgend in Glamorganshire on the
6th of June 1750 and died May the 4th 1833 aged 82.
(Inscription on William Morgan's tomb in
St Mary's churchyard, Hornsey)

William's chest tomb in St Mary's churchyard, Hornsey, is just off the busy High Street, but you barely notice the hum of traffic. It's a peaceful green corner, though not as quiet as when William chose the place for his family burials and Hornsey was still little more than a country village. The original church, a small low building with a squat little tower, had been enlarged just before William's death to accommodate the growing population, and the tower increased in height to balance the extended nave. The church has gone now, demolished in 1927, but the tower remains and you can see the change in brickwork where the extension begins. Shrubs and flowers have been planted on the site of the old nave to create a Garden of Remembrance and in summer lavender and roses scent the air.

Many of the graves have been cleared away but the Morgan tomb is one of the few which remain. It's in a patch of rough grass and, in the same section, there is another – larger and rather grander – chest tomb surrounded by railings. Samuel Rogers is buried just a few steps away from his lifelong friend. Rogers lived to the age of ninety-two and he was nearly eighty-seven when, on the death of Wordsworth,

FIGURE 40 The Morgan family tomb at St Mary's, Hornsey.

he was invited to become the Poet Laureate. Prince Albert assured him that 'the practice (at all times objectionable) of exacting lauda-tory Odes' was no longer expected.[1] The post carried no duties; it was a compliment and an honour. Rogers agonised. His reply is as sad to read as it must have been to write. '[A]fter long deliberation and many conflicts within me', he wrote to Prince Albert, 'I am come, but with great reluctance, to the resolution that I must decline the offer, but sub-scribing myself, with a gratitude that will not go but with the last beat of my heart.'[2] Tennyson was chosen instead, but not before the Prime Minister, Lord John Russell, had written to Rogers to check Tennyson's moral suitability. Rogers generously gave Tennyson not only his support but also the loan of his court dress. As with Wordsworth, seven years previously, Tennyson had to be squeezed into the outfit.

By the end of the century there were eight members of the Morgan family buried at Hornsey: William and his wife, Susanna; his daughter, Susan; and three of his sons: William, John and Arthur; also Arthur's wife, Louisa, and Frances, the daughter of William (junior). Sarah's death is recorded on the inscriptions and the bald facts about her re-interment in the Travers's tomb at Hendon. William's fourth son, Cadogan, is also named and the fact that his grave is at Coity.

Cadogan remains a rather shadowy figure in the family tree. He was admitted to the Middle Temple in 1823, called to the Bar in 1828 and, for some years, practised as a barrister in London. He married at the age of fifty-seven and moved to Bridgend, where his one surviving daughter was born.

Arthur was successful as Actuary at the Equitable though, like William, he had on occasions a tough time dealing with the demands of members.[3] In 1835 he was elected as a Fellow of the Royal Society, principally for his work on tabulating the Equitable's mortality figures. Interestingly, one of his sponsors was Benjamin Travers. Perhaps this suggests some thaw in the family feud or perhaps it was a strictly professional matter – by this time Travers was one of the directors at the Equitable. Arthur and Louisa had no children. Did Arthur 'grow up' and become solemn and serious, or did he retain his sense of fun and become a jolly uncle to John's family?

John and Anne had eight children, seven of whom lived into adulthood though only three married and had children. His son, another William, had a large family and, among descendants, the Morgan name lives on to this day. Elsewhere the name is lost as the direct line is through two of his daughters: Mary Susan (my great-grandmother), who married a surgeon in the Indian Army, and Annie, who married Dr Alexander Waugh – a name made famous by their many literary descendants.

John became a distinguished surgeon. According to his obituarist, he 'evinced a wholesome tardiness in resorting to the knife, yet he was a bold, unflinching, and enterprising operator'.[4] He was interested in comparative anatomy, keeping, in the backyard of his consulting rooms, some female kangaroos.[5] On one occasion he dissected an elephant which had been shot 'in a rabid state'[6] at the Exeter Exchange – an extraordinary building in the Strand which housed a menagerie of wild animals. Together with John Bright, Thomas Addison and Thomas Hodgkins (all of whose names are now used for the diseases they identified), he investigated the possibility of treating the terrible convulsive

spasms of tetanus and rabies with the paralysing poisons used in hunt-
ing by South American Indians. But John's chief work was ophthalmol-
ogy; he was a key mover in establishing the first eye infirmary at Guy's
Hospital 'affording relief to thousands'.[7]

John's death, at the age of only fifty, left Anne with seven chil
dren aged between fifteen and three. There followed years of straitened
means in 'gloomy houses in Clifton [Bristol], in a cottage on the little
family estate in the Rhondda valley, or in a prim, "genteel" residence
in Poole'.[8]

Sarah Travers, William's ward and granddaughter, lived at Stamford
Hill until, in her late forties, she moved to Norwood. She was an affectionate and much-loved member of the family and, when she died
in her ninetieth year, her will was full of thoughtful bequests to her
half-brothers and sisters and their children; her nephews, nieces and
cousins.

As for William's house at Stamford Hill, it remained in the family
until Arthur died in 1870. At the beginning of the twentieth century
it was bought by the YMCA and opened as a hostel in 1909 by the
Lord Mayor of London in front of 'a large gathering of influential
people'.[9] Later Nathan Rothschild's house next door was also bought,
and the two properties provided spacious grounds 'so well wooded
that ... one can hardly realize that it is in London'.[10] A gymnasium
was built in the gardens, also tennis courts. It was used as a place of
convalescence during the First World War and photographs show young
men well enough to take part in hay making. The house was eventually
demolished in the 1960s and a modern block of flats now stands on
the site.

FIGURE 41 Stamford Hill, *c.*1918, when it was a YMCA hostel.
These pictures show young men and women recovering
health after the First World War.
*(Cadbury Research Library, Special Collections, University
of Birmingham, YMCA/K/1/12/143)*

EPILOGUE

It's 2016. I take my ten-year-old grandson to At-Bristol, the city's interactive science centre.[1] Jack has been before and he is keen to show me his favourite exhibits. We squeeze handfuls of iron filings which cling together and bridge the gap between two magnets. Then we lift them beyond the magnets' grip and feel them loosen and flow through our fingers. We create patterns in sand on a spinning disc and, in the shop, we look at plasma lamps.

I think of William Morgan. At-Bristol is everything which would please him: curiosity nurtured, experiment encouraged and, everywhere, hands-on learning about our world. William, the wise and loving father who let his five-year-old son sharpen his tools, would rejoice to see this distant grandson looking, feeling, testing, discovering. Two hundred years and seven generations separate William and Jack, but a spirit of inquiry and wonder shrinks the years and unites them now and into the future.

NOTES

Chapter 1

1. Williams, *A Welsh Family*, p. 39.
2. See Williams, *A Welsh Family*, pp. 38–9.
3. Cowbridge Grammar School was owned by Jesus College Oxford. For details of this see Iolo Davies, *A Certaine Schoole, A History of the Grammar School at Cowbridge, Glamorgan* (Cowbridge: D. Brown and Sons Ltd, 1967).
4. Williams, *A Welsh Family*, p. 36.
5. Williams, *A Welsh Family*, p. 43.
6. See Williams, *A Welsh Family*, ch. 3, for an account of the marriage and family life of Dr William Morgan and Sarah Price.
7. Williams, *A Welsh Family*, p. 24. The modern Welsh spelling is 'Hoga fy mwyell'.
8. In Williams, *A Welsh Family*, pp. 23 and 45, Caroline Williams spells the name Tyl yr Coch. I am grateful to Gareth Roberts and Richard Morgan for giving me the correct spelling.
9. H. J. Randall, *Bridgend: The Story of a Market Town* (Newport: R. H. Johns, 1955), p. 140.
10. Williams, *A Welsh Family*, p. 28.
11. For an examination of journey times and travel conditions see Philip S. Bagwell, *The Transport Revolution from 1770* (London: B. T. Batsford Ltd, 1974), pp. 34–60.
12. John Rocque, with introductory notes by Ralph Hyde, *The A to Z of Georgian London*, London Topographical Society Publication No. 126 (Kent: Harry Margary, in association with Guildhall Library, London, 1982).
13. Samuel Rogers, *Table-Talk & Recollections*, selected by Sir Christopher Ricks (London: Notting Hill Editions Ltd, 2011), p. viii. Rogers's first editor was a contemporary, Alexander Dyce, whose edition was followed three years later by a more concise volume edited by his nephew, William Sharpe. Ricks selects from the work of both editors, but mainly from Sharpe, p. 2.
14. Work to clean St Paul's began in 1996 and, when cleaning was finished in 2011, Martin Stancliffe, Surveyor to the Fabric, said it probably looked 'better than at any time since its completion in 1711'. Available at *https://www.stpauls.co.uk/documents/Press%20releases/Completion%20of%20programme%20of%20cleaning%20and%20repair-%20press%20release.pdf*. Accessed 2016.

15. I am very grateful to Alex Allardyce, author of *The Village that Changed the World, A History of Newington Green* (London: Newington Green Action Group, 2008), for showing me round Newington Green and arranging my visit to number 54.

Chapter 2

1. Rogers, *Table-Talk*, p. viii.
2. Fanny Kemble, *Records of Later Life* (New York: H. Holt and Co., 1882), p. 65.
3. P. W. Clayden, *The Early Life of Samuel Rogers* (London: Smith, Elder and Co., 1887), p. 8. Clayden also records Samuel Rogers's anecdotes about Richard Price, pp. 9–10.
4. Williams, *A Welsh Family*, p. 34.
5. William Morgan, *Memoirs of the Life of the Rev. Richard Price* (London: R. Hunter and R. Rees, 1815; repr. ed. D. O. Thomas, in *Enlightenment and Dissent*, 22 (2003)), pp. 42–5. All subsequent references to the *Memoirs* are to the Thomas-edited version.
6. Morgan, *Memoirs of the Life of Richard Price*, ed. Thomas, p. 43.
7. Morgan, *Memoirs of the Life of Richard Price*, ed. Thomas, p. 8.
8. For a discussion of the argument between Price and Hume see Bernard Peach, 'On What Point did Richard Price Convince David Hume of a Mistake? With a Note by Henri Laboucheix', in *The Price–Priestley Newsletter*, 2 (1978), 76–81.
9. For a readable account of the history of apothecaries see W. S. C. Copeman, *Apothecaries of London: A History 1617–1967* (London: Pergamon Press, 1967), pp. 15–22.
10. Williams, *A Welsh Family*, p. 37.
11. Williams, *A Welsh Family*, p. 38.
12. Liza Picard, *Dr Johnson's London* (London: Phoenix Press, 2001), pp. 99–100.
13. Picard, *Dr Johnson's London*, p. 89.
14. Williams, *A Welsh Family*, p. 38.
15. Clayden, *The Early Life of Samuel Rogers*, p. 9.
16. Tom Tucker, *Bolt of Fate: Benjamin Franklin and his Electric Kite Hoax* (Stroud: Sutton Publishing Ltd, 2004), p. xvi.
17. See Tucker, *Bolt of Fate*, chs 1–4, for details of some of the electrical displays.
18. 'Letter from Houlton Harries (Surgeon to the Forces, Barbados) addressed to the Revd Dr Price', Royal Society/Letters&Papers/8/133, hereafter RS/L&P.
19. Paper by Riddleston, presented by Dr Price, RS/L&P/8/134.
20. 'Letter from James Nooth (Surgeon at Dorchester) addressed to Sir John Smith', RS/L&P/8/135.
21. 'Letter from J. Lane sent to Dr Watson', RS/L&P/8/132.
22. RS/L&P/8/133.
23. RS/L&P/8/132.

Chapter 3

1. Tobias Smollett, *The Expedition of Humphry Clinker*, Oxford World's Classics edn (Oxford: Oxford University Press, 1984), p. 87.
2. Williams, *A Welsh Family*, p. 35.

3. Williams, *A Welsh Family*, p. 34.

4. Williams, *A Welsh Family*, p. 42.

5. Williams, *A Welsh Family*, p. 35.

6. Williams, *A Welsh Family*, p. 51. See also Maurice Ogborn, *Equitable Assurances: The Story of Life Assurance in the Experience of the Equitable Life Assurance Society 1762–1962* (London: George Allen and Unwin Ltd, 1962), pp. 101–2.

7. The appointment is recorded in the Minutes of the Summoned Court of the Society of Equitable Assurances for 21 February 1774, EL/1/3/2. All the Equitable's documents are now held at the Institute of Actuaries, Staple Inn.

8. Bond of Security, Glamorgan Archives, Merthyr Mawr Estate Collection, DMM/PR/165/1.

9. Letter of 18 March 1775 to Samuel Price, Glamorgan Archives, Merthyr Mawr Estate Collection, DMM/PR/165/2.

10. The directors' comment is recorded in the Minutes of the Summoned Court for 27 January 1775, EL/1/2/8.

11. The appointment is recorded in the Minutes of the Summoned Court for 16 February 1775, EL/1/2/8.

12. Richard Price, 'Observations on the Expectations of Lives, The Increase of Mankind, The Influence of Great Towns on Population, and Particularly the State of London, with Respect to Healthfulness and Number of Inhabitants. Communicated to the Royal Society, April 27, 1769. In a Letter from Mr Richard Price F.R.S. to Benjamin Franklin, Esq.; LL.D. and F.R.S.', *Philosophical Transactions of the Royal Society*, 59 (1769). Also Richard Price, *Observations on Reversionary Payments; on Schemes for Providing Annuities for Widows, and for Persons in Old Age; on the Method of Calculating the Values of Assurances on Lives; and on the National Debt* (London: Cadell, 1771).

13. Ogborn, *Equitable Assurances*, p. 104.

14. Craig Turnbull, *A History of British Actuarial Thought* (London: Palgrave Macmillan, 2017), pp. 66–7.

15. Ogborn, *Equitable Assurances*, p. 105.

16. William Morgan, *The Doctrine of Annuities and Assurances on Lives and Survivorships, stated and explained. To which is added an Introduction, addressed to the Society; also An Essay on the present State of Population in England and Wales, by the Reverend Dr Price* (London: T. Cadell, 1779, repr. Gale Eighteenth Century Collections Online Print Edition, hereafter Gale ECCO).

17. Morgan, *Doctrine of Annuities and Assurances*, p. v.

18. Morgan, *Doctrine of Annuities and Assurances*, p. vi.

19. William Pitt the Elder, Earl of Chatham (1708–78), was prime minister from 1766 to 1768.

20. Copeman, *A History of the Worshipful Society of Apothecaries of London*, pp. 32–3.

21. Letter of 3 January 1803 to John Price, Glamorgan Archives, Merthyr Mawr Estate Collection, DMM/PR/52/4.

22. See Ogborn, *Equitable Assurances*, p. 52, for details about applications for policies.

23. All the names of the policyholders are recorded in the Minutes of the meetings of

the Weekly Court of Directors of the Society of Equitable Assurances. I am grateful to David Raymont, Librarian at the Institute of Actuaries, for his help in finding the records of these policies.

24. William Strange, *Sketches of Her Majesty's Household interspersed with Historical Notes, Political Comments, and Critical Remarks* (London: William Strange, 1848), pp. 91–2. Strange identifies the duty of the Table Deckers as being 'to superintend the arrangement of her Majesty's table; placing everything in perfect order previously to the dinner being served', and notes that Queen Victoria had three Table Deckers, with an assistant and a wax-fitter.

25. Quoted by M. Dorothy George in *London Life in the Eighteenth Century* (London: Peregrine Books, 1966), p. 339, n.63 to ch. 2.

26. César de Saussure, *A Foreign View of England in the Reigns of George I and George II* (London: John Murray, 1902), p. 81.

27. Sophie van la Roche, 'Diary (1786)', in Rick Allen (ed.), *The Moving Pageant: a Literary Sourcebook on London Street-life, 1700–1914* (London and New York: Routledge, 1998), p. 76.

28. van la Roche, 'Diary', p. 76.

29. Quoted by Peter Ackroyd in *The History of England, Vol. IV: Revolution* (London: Macmillan, 2016), p. 210.

Chapter 4

1. *The Gentleman's Magazine*, 50 (1780), 312.

2. J. Paul de Castro, *The Gordon Riots* (Oxford: Oxford University Press, 1926), pp. 29–31.

3. See de Castro, *The Gordon Riots*, p. 37.

4. Picard, *Dr Johnson's London*, p. 295.

5. Letter from Lady Anne Erskine to her brother, the eleventh Earl of Buchan. See de Castro, *The Gordon Riots*, pp. 106–7.

6. For a full discussion of these images see Ian Haywood, 'A metropolis in flames and a nation in ruins: the Gordon riots as sublime spectacle', in Ian Haywood and John Seed (eds), *The Gordon Riots: Politics, Culture and Insurrection in Late Eighteenth-Century Britain* (Cambridge: Cambridge University Press, 2012), pp. 119–29.

7. Charles Dickens, *Barnaby Rudge* (London: Everyman's Library, 2005 edn), pp. 556–8.

8. Williams, *A Welsh Family*, p. 67.

9. Haywood and Seed (eds), *The Gordon Riots*, p. 7.

10. Matthew White, '"For the safety of the city": the geography and social politics of public execution after the Gordon riots', in Haywood and Seed (eds), *The Gordon Riots*, pp. 204–14.

11. Matthew White, '"For the safety of the city"', in Haywood and Seed (eds), *The Gordon Riots*, p. 219.

12. P. W. Clayden, *The Early Life of Samuel Rogers*, p. 38. Rogers's recollection (or perhaps Clayden's account) is not entirely reliable, since he claims they were on their way to

Tyburn. None of the hangings took place at Tyburn. What is not in doubt is Rogers's compassion for the girls in coloured dresses.

13. Quoted in Haywood and Seed (eds), *The Gordon Riots*, p. 2.

14. See chapter 13, 'The Trumpet of Liberty'.

Chapter 5

1. Letter of 3 October 1775 from Franklin to Price, in D. O. Thomas and Bernard Peach (eds), *The Correspondence of Richard Price*, vol. 1 (Cardiff: University of Wales Press, 1983), p. 229.

2. See D. O. Thomas, 'George Cadogan Morgan (1754–1798)', in *The Price–Priestley Newsletter*, 3 (1789), 53. For a discussion of Price and Unitarianism see Roland Thomas, *Richard Price, Philosopher and Apostle of Liberty* (Oxford: Oxford University Press, 1924), pp. 65–6. Also R. K. Webb, 'Price among the Unitarians', *Enlightenment & Dissent*, 19 (2000), 147–70.

3. See Roland Thomas, *Richard Price, Philosopher and Apostle of Liberty* (Oxford: Oxford University Press, 1924), pp. 21–2.

4. Williams, *A Welsh Family*, p. 53.

5. See Williams, *A Welsh Family*, pp. 56–7.

6. Allardyce, *The Village that Changed the World*, Introduction. See also pp. 18–26 for an account of the people connected with Newington Green in the eighteenth century.

7. See *The Christian Reformer, or, Unitarian Magazine and Review* (London: Edward T. Whitfield, 1856), vol. XII, CXXXVI, 197, for a list of Price's visitors. Adam Smith and Thomas Jefferson corresponded with Price but may not have visited Newington Green.

8. The Earl of Shelburne became the Marquis of Lansdowne in 1784.

9. For a discussion of this change of world view, see Richard Holmes, *The Age of Wonder: How the Romantic Generation Discovered the Beauty and Terror of Science* (London: Harper Press, 2008), pp. 245–7.

10. See Frame, *Liberty's Apostle*, p. 85–6, for an entertaining discussion of some of the societies of the period.

11. See *OED*, 'ornament, 2 c, A person who enhances or adds distinction to his or her sphere, time, etc.', with examples from 1422–1989.

12. Dr Márcia Balisciano, *Benjamin Franklin and Public History*, Introduction. Available at *http://www.benjaminfranklinhouse.org/site/sections/news/MBArticle.pdf*. Accessed 2015.

13. See, for example, Tucker, *Bolt of Fate*, pp. 135–56. Tucker finds problems with every detail of the reported experiment, from the materials used for the kite to the construction of the shed. He analyses Franklin's account, published first in the *Pennsylvanian Gazette* and subsequently in the *Philosophical Transactions of the Royal Society*. There are no specifics in Franklin's description: 'No witness, no location named, no year, no month, no day, no hour supplied.' Tucker also points out that Priestley's account of his friend's report, written fifteen years later, includes a number of additional details, such as Franklin's son being present.

14. I. Bernard Cohen, *Benjamin Franklin's Science* (Cambridge, MA: Harvard University Press, 1996), p. 157.

15. Joseph Priestley, *The History and Present State of Electricity, with Original Experiments* (London: J. Dodsley, J. Johnson, B. Davenport and T. Cadell, 1767), p.159.

16. Recorded in the *Journal Book of the Royal Society*, 22 (1751–4), 410, 30 November 1753 (Royal Society Archive, JBO/22).

17. George Cadogan Morgan, *Lectures in Electricity*, vol. 1 (Norwich: J. March, 1794, repr. Gale ECCO), p. xxxviii.

18. These superstitious beliefs are discussed in various publications. See, for example, Simon Schafer, 'Charged Atmospheres: Promethean Science and the Royal Society', in Bill Bryson (ed.), *Seeing Further: The Story of Science and the Royal Society* (London: Harper Press, 2010), p. 143; also Tucker, *Bolt of Fate*, p. 127.

19. Quoted by Tucker, *Bolt of Fate*, p. 127.

20. Morgan, *Memoirs of the Life of Richard Price*, ed. Thomas, pp. 24–5.

21. Morgan, *Memoirs of the Life of Richard Price*, ed. Thomas, p. 26.

22. Clayden, *The Early Life of Samuel Rogers*, p. 33.

23. Morgan, *Memoirs of the Life of Richard Price*, ed. Thomas, p. 28.

24. For details about the pamphlet and its sales see Frame, *Liberty's Apostle*, pp. 110–11.

25. Clayden, *The Early Life of Samuel Rogers*, p. 34. A carman is a carter, a carrier.

26. See the letter of 23 March 1776 by Price to William Rix, Town Clerk of the City of London, in Peach and Thomas (eds), *The Correspondence of Richard Price*, vol. I, p. 243, n.2.

27. Morgan, *Memoirs of Richard Price*, ed. Thomas, p. 29.

28. Williams, *A Welsh Family*, p. 57.

29. Letter of 26 October 1779 from Price to Baron J. D. van der Capellen, in D. O. Thomas (ed.), *Correspondence of Richard Price*, vol. II (Cardiff: University of Wales Press, 1991), p. 55.

30. Quoted by William Hague in *William Pitt the Younger* (London: HarperCollins, 2004), p. 76.

31. For details of the peace treaty see Frame, *Liberty's Apostle*, pp. 155–6.

Chapter 6

1. Sir Humphry Davy, 'On the Electrical Phenomena Exhibited in Vacuo', in *Philosophical Transactions of the Royal Society*, 112 (1822), 64.

2. See Holmes, *The Age of Wonder*, pp. 75–7, for details of Herschel's life in Bath, and pp. 96–106 for details of his discovery, whilst in Bath, of Uranus.

3. See letter from Price to James Bowdoin, 25 October 1785, in Peach and Thomas (eds), *The Correspondence of Richard Price*, vol. II, p. 315.

4. Letter from Price to Franklin, 6 April 1784, in Peach and Thomas (eds), *The Correspondence of Richard Price*, vol. II, p. 214.

5. William Morgan, 'Electrical Experiments Made in Order to Ascertain the Non-Conducting Power of a Perfect Vacuum', in *Philosophical Transactions of the Royal Society*, 75 (1785), 272–8.

6. *Philosophical Transactions of the Royal Society*, 75 (1785), 273.

7. Sir Humphry Davy, 'On the Electrical Phenomena Exhibited in Vacuo', in
   Philosophical Transactions of the Royal Society, 112 (1822), 64–75.
8. Sir Humphry Davy, 'On the Electrical Phenomena Exhibited in Vacuo', in
   Philosophical Transactions of the Royal Society, 112 (1822), 65.
9. Michael Faraday, 'Experimental Researches in Electricity – Thirteenth Series', in
   Philosophical Transactions of the Royal Society, 128 (1838), 154–5.

Chapter 7

1. Quoted in the exhibition catalogue of *The Royal Society: 350 Years of Science* (London:
   The Royal Society, 2010), p. 15.
2. The policies are recorded in the Minutes of the Weekly Court of Directors of the
   Society of Equitable Assurances (WCoD):
   1. James, Duke of Chandos, Policy No. 6793, Minutes of the Weekly Court of
   Directors of the Society of Equitable Assurances (WCoD), 25 January 1786
   [EL/1/2/14] (Alexander Ludens insured the life of the Duke of Chandos, aged 55,
   for one year).
   2. Henry Frederick, Duke of Cumberland, Policy No. 4025, WCoD, 19 January 1780
   [EL/1/2/10] (Henry De La Douespe insured the life of the Duke of Cumberland,
   aged 35, for the continuance of his life).
   3. William, Duke of Devonshire, Policy No. 8099, WCoD, 7 November 1787
   [EL/1/2/16] (Edward Barnett insured the life of the Duke of Devonshire, aged 39,
   for seven years).
   4. Henry, Earl of Clanrickarde, Policy No. 7083, WCoD, 19 July 1786 [EL/1/2/15]
   (George Fillingham insured the life of the Earl of Clanrickarde, aged 47, for seven
   years).
   5. William, Earl of Shelburne, Policy No. 3598, WCoD, 8 July 1778 [EL/1/2/9]
   (Joseph Sparkes insured the life of the Earl of Shelburne, aged 41, for the
   continuance of his life).
   6. Earl of Derby, Policy No. 4738, WCoD, 3 April 1782 [EL1/2/10] (Edward Moore
   insured the life of the Earl of Derby, aged 31, for seven years).
   7. Arthur, Earl of Donegall, Policy No. 6370, WCoD, 25 May 1785 [EL1/2/14]
   (John Martin insured the life of the Earl of Donegall, aged 47, for the continuance
   of his life).
   8. Lord Melbourne, Policy No. 5591, WCoD, 11 February 1784 [EL/1/2/12]
   (Humphrey Owen insured the life of Lord Melbourne, aged 44, for one year).
3. Mary Wells (Junior), Policy No. 5621, WCoD, 25 February 1784 [EL1/2/13]
   (Samuel Smith insured the lives of Mary Wells, junior, aged 51, and Lydia Wells,
   aged 39, beyond the life of mother Mary Wells).
4. See chapter 3, p. 32: insuring the life of another in the hope of a reward such as
   inheritance was illegal after 1774.
5. Policy No. 5054, WCoD, 29 January 1783 [EL/1/2/11].
6. A twenty-first-century actuary is just as likely to be a woman as a man.
7. Later called the Laudable Society of Annuitants.

8. For a full list see D. R. Bellhouse, *Leases for Lives: Life Contingent Contracts and the Emergence of Actuarial Science in Eighteenth-Century England* (Cambridge: Cambridge University Press, 2017), p. 137.

9. See John V. Tucker, 'Richard Price and the History of Science', *Transactions of the Honourable Society of Cymmrodorion*, 23 (2017), 84.

10. C. G. Lewin, *Pensions and Insurance before 1800: A Social History* (East Lothian, Scotland: Tuckwell Press Ltd, 2003), pp. 292–300.

11. See Craig Turnbull, *A History of British Actuarial Thought* (London: Palgrave Macmillan, 2017), pp. 75–81.

12. 'An Essay towards Solving a Problem in the Doctrine of Chances, by the late Rev. Mr Bayes, F.R.S., communicated by Mr Price, in a letter to John Canton, A.M. F.R.S.', *Philosophical Transactions of the Royal Society*, 53 (1763), 370–418.

13. Bryson (ed.), *Seeing Further*, p. 3.

14. For an explanation of Bayes's Theorem see Turnbull, *A History of British Actuarial Thought*, pp. 38–43, and for a full discussion of the many applications of Bayes's Rule see Sharon Bertsch McGrayne, *The Theory That Would Not Die* (New Haven and London: Yale University Press, 2011), pp. 213–32.

15. For a more detailed analysis of the application of Bayes's Rule to insurance see Tucker, 'Richard Price and the History of Science', 77–86.

16. 'Observations on the expectations of lives, in the increase of mankind, the influence of great towns on population, and particularly the state of London with respect to healthfulness and number of inhabitants', in *Philosophical Transactions of the Royal Society*, 59 (1769).

17. Richard Price, *Observations on Reversionary Payments; on schemes for providing annuities for widows, and for persons in old age*, seventh edition rearranged and enlarged by William Morgan (London: T. Cadell and W. Davies, 1812).

18. 'On the Probabilities of Survivorship between Two Persons of Any Given Ages, and the Method of Determining the Values of Reversions Depending on Those Survivorships', Communicated by the Rev. Richard Price, D.D.F.R.S., *Philosophical Transactions of the Royal Society*, 78 (1788), 331–49.

19. 'On the Method of Determining, from the Real Probabilities of Life, the Value of a Contingent Reversion in Which Three Lives are Involved in the Survivorship', Communicated by the Rev. Richard Price, D.D.F.R.S., *Philosophical Transactions of the Royal Society*, 79 (1789), 40–54.

20. See Lewin, *Pensions and Insurance before 1800*, for a readable and detailed account of the history of life assurances from its earliest stages.

21. See Turnbull, *A History of Actuarial Thought*, p. 50.

22. See Ogborn, *Equitable Assurances*, pp. 26 and 257.

23. He was the first to state that, if a life office were to be established on proper scientific principles, then the calculated premium for a life policy must be calculated (from the mortality table and the interest rate) so as to depend on the age of the person taking the life policy out. Furthermore, Dodson showed how the accumulated premiums, together with interest thereon, would build up in time but would be diminished later on in time by the death claims that would need to be paid. Scientific principles determined that

the discounted value of the premiums plus expenses plus a safety margin (to guard against deteriorating mortality or falling interest rates or rising expenses or other unexpected events, like rising taxes) would equal the discounted value of claims.

24. Morgan, 'On the Probabilities of Survivorship between any Two Persons of Any given Age and the Method of Determining the Values of Reversions Depending on Those Survivorships', in *Philosophical Transactions of the Royal Society*, 78 (1788), 331–2.

25. Morgan, 'On the Probabilities of Survivorship between any Two Persons of Any given Age and the Method of Determining the Values of Reversions Depending on Those Survivorships', in *Philosophical Transactions of the Royal Society*, 78 (1788), 332.

26. *The Times*, Tuesday 2 May 1933.

27. Elderton, 'William Morgan', in *Transactions of the Faculty of Actuaries*, 14 (1931–4), 15.

28. William Morgan, *A View of the Rise and Progress of the Equitable Society, and of the causes which have contributed to its success: to which are added remarks on some of the late misrepresentations respecting the rules and practice of the Society* (London, Longman, Rees, Orme, Brown and Green, 1829, repr. Gale The Making of Modern Law Print Editions, hereafter MOML), note on p. 16.

29. See note in Elderton, 'William Morgan', in *Transactions of the Faculty of Actuaries*, 14 (1931–4), 15. On 15 March 1825, before the Committee on Friendly Societies, William said, 'You have, in the Act of Parliament, made use of the words, two actuaries, or professional men. The people in this country have taken upon themselves this name, I have had cases sent me, of club rules and orders, provided by people who call themselves actuaries, who are nothing but schoolmasters and accountants, and some of the tables are exceedingly wrong. The mischief is now very great.'

Chapter 8

1. William Morgan, *An Examination of Dr Crawford's Theory of Heat and Combustion* (London: T. Cadell, 1781, repr. Gale ECCO), p. 1.

2. Adair Crawford, *Experiments and Observations on Animal Heat, and the Inflammation of Combustible Bodies; being an attempt to resolve these phenomena into a general law of nature* (London: J. Murray and J. Sewell, 1779).

3. Morgan, *An Examination of Dr Crawford's Theory of Heat and Combustion*, p. 3.

4. Morgan, *An Examination of Dr Crawford's Theory of Heat and Combustion*, p. 60.

5. Morgan, *An Examination of Dr Crawford's Theory of Heat and Combustion*, p. 70.

6. Morgan, *An Examination of Dr Crawford's Theory of Heat and Combustion*, p. 70.

7. All the words of this dispute are taken from articles in the *Gentleman's Magazine*, one by 'X. Y. Z.' in vol. 58 (1788), 895–9, and two by 'A. B.' in vol. 59 (1789), 129–32, 219–21. I have used some artistic licence in presenting the exchange in a selection of alternating short comments and answers.

8. Letter of 17 April 1790 from Richard Price to Sir Joseph Banks, in W. Bernard Peach (ed.), *The Correspondence of Richard Price* (Durham, NC: Duke University Press; Cardiff: University of Wales Press, 1994), vol. III, p. 279, n.1. The 'celebrated dissenting patriot' is Richard Price, and William Morgan was awarded the Copley Medal in 1789 (see chapter 7).

9. Letter of 17 April 1790 from Richard Price to Sir Joseph Banks, p. 280.
10. Letter of 27 April 1790 from Price to Sir Joseph Banks, in Peach (ed.), *The Correspondence of Richard* Price, vol. III, pp. 287–8.
11. See Paul Frame, *Liberty's Apostle*, p. 55, for a discussion of Maseres's scheme.
12. The Linnean Society of London, named after the Swedish naturalist Carl Linnaeus (1707–78), is the world's oldest active biological society, documenting flora and fauna and maintaining botanical, zoological and library collections.
13. See Russell McCormmach, *Speculative Truth: Henry Cavendish, natural philosophy and the rise of modern theoretical science* (Oxford, New York: Oxford University Press, 2004).
14. Available at *https://collections.royalsociety.org/DServe.exe?dsqIni=Dserve.ini&dsqApp= Archive&dsqDb=Catalog&dsqSearch=RefNo=='EC%2F1790%2F05'&dsqCmd=Show.tcl*. Accessed 2012.
15. The full list of signatories is: Henry Cavendish, Charles Blagden, Francis Maseres, John Paradise, Nevil Maskelyne, Richard Price, James Edward Smith, Timothy Lane, Andrew Kippis, Abraham Rees, Thomas Emlyn, Henry Partridge.
16. James Gleick, 'At the Beginning: More Things in Heaven and Earth', in Bryson (ed.), *Seeing Further*, p. 18.
17. Thomas Sprat, *History of the Royal Society* (1667), quoted in the exhibition catalogue of *The Royal Society: 350 Years of Science* (London: The Royal Society, 2010), p. 10.

Chapter 9

1. Reputed to be James I's exclamation when he arrived at Stamford Hill, having travelled from Scotland to take possession of the crown. See M. Bernstein, *Stamford Hill and the Jews before 1915* (London: MSB Publications, 1976), p. 11.
2. William Palin Elderton, 'Some Family Connections of William Morgan (1750–1833) F.R.S.', in *The Genealogists' Magazine*, 12/10 (June 1957), 332.
3. Williams, *A Welsh Family*, p. 66.
4. Ogborn, *Equitable Assurances*, p. 197.
5. See Bellhouse, *Leases for Lives*, pp. 199–200 and 202 for details of this and other consultancy work.
6. Rena Gardiner, *The Story of Saint Bartholomew the Great* (London: Select Typesetters, 2005), p. 29.
7. Susanna had five siblings, three sisters and two brothers. One of her elder sisters, Hannah (1748–1817) was married to Samuel Crompton (dates unknown).
8. St Bartholomew the Great Register of Marriages and Banns (1783–7), held at the London Metropolitan Archive, P69/BAT3/A/004/MS06779/003.
9. Williams, *A Welsh Family*, pp. 68–9.
10. Williams, *A Welsh Family*, p. 71.
11. St Illtud's church records.
12. George Borrow, *Lavengro*, quoted by C. B. Jewson in *The Jacobin City: A Portrait of Norwich in its Reaction to the French Revolution 1788–1802* (Glasgow and London: Blackie and Son, 1975), p. 1.
13. Williams, *A Welsh Family*, p. 82.

14. Morgan, *Memoirs of Richard Price*, ed. Thomas, p. 58.
15. Williams, *A Welsh Family*, p. 85.
16. Peach (ed.), Letter of 27 September 1786 from Price to George Cadogan Morgan, in *The Correspondence of Richard Price*, vol. III, pp. 61–3.
17. The Marquis of Lansdowne, earlier the Earl of Shelburne, became marquis in 1784. For a discussion of Price's friendship with the Marquis of Lansdowne see Frame, *Liberty's Apostle*, pp. 63–4.
18. Williams, *A Welsh Family*, p. 88.
19. Williams, *A Welsh Family*, p. 89.
20. I am grateful to Rosemary Harden of the Fashion Museum, Bath, for examining the photograph and, by considering the women's clothes and hairstyles, also the men's jackets, suggesting this date.
21. Williams, *A Welsh Family*, p. 88.
22. Williams, *A Welsh Family*, p. 86. William Morgan always considered it 'a grave reproach' to the (Gravel Pit) congregation that they did not choose George to succeed Price.

Chapter 10

1. Richard Price, 'A Discourse on the Love of our Country', in D. O. Thomas (ed.), *Price: Political Writings* (Cambridge: Cambridge University Press, 1991), p. 196.
2. Letter of Price to John Howard, 31 January 1789, in Peach (ed.), *The Correspondence of Richard Price*, vol. III, p. 205.
3. For a concise discussion of the symptoms and twentieth-century diagnosis of the disease see Jane Ridley, *Bertie: A Life of Edward VII* (London: Chatto and Windus, 2012), pp. 7–8.
4. See Hague, *Pitt the Younger*, p. 236.
5. Letter of Price to John Howard, 31 January 1789, in Peach (ed.), *The Correspondence of Richard Price*, vol. III, p. 205.
6. See Hague, *Pitt the Younger*, pp. 265–7.
7. Peach (ed.), Letters from Price to Thomas Jefferson in *The Correspondence of Richard Price*, vol. III, 8 January, pp. 195–9; 12 May, pp. 218–19; 19 May, pp. 223–5; and 12 July, pp. 231–7.
8. See William Doyle, *The Oxford History of the French Revolution* (Oxford and New York: Oxford University Press, 1990), pp. 67–9.
9. Doyle, *The Oxford History of the French Revolution*, p. 105.
10. Morgan, *Memoirs of the Life of Richard Price*, ed. Thomas, pp. 74–5.
11. Morgan, *Memoirs of the Life of Richard Price*, ed. Thomas, p. 75.
12. Morgan, *Memoirs of the Life of Richard Price*, ed. Thomas, p. 76.
13. Morgan and Morgan, *Travels in Revolutionary France*, ed. Constantine and Frame, p. xi.
14. For a discussion of this see Morgan and Morgan, *Travels in Revolutionary France*, ed. Constantine and Frame, p. 11.
15. Lady Elizabeth Rigby Eastlake (ed.), *Dr Rigby's Letters from France &c in 1789* (London: Longmans, Green and Co., 1880; repr. Lightning Source UK Ltd),

p. 7. Rigby mentions 'letters to Mirabeau, Target and some others of the popular characters of the Assembly'.

16. Letter of 1 July 1789 from Price to Mathon de la Cour, in Peach (ed.), *The Correspondence of Richard Price*, vol. III, p. 229.

17. Letter of 25 September 1784 from Price to William Temple Franklin, in Thomas (ed.), *The Correspondence of Richard Price*, vol. II, p. 227.

18. Letter of 2 July 1789 from Price to Count Mirabeau, in Peach (ed.), *The Correspondence of Richard Price*, vol. III, p. 229.

19. Eastlake (ed.), *Dr Rigby's Letters*, p. 6.

20. Eastlake (ed.), *Dr Rigby's Letters*, p. 2.

21. Morgan and Morgan, *Travels in Revolutionary France*, ed. Constantine and Frame, p. 43.

22. Morgan and Morgan, *Travels in Revolutionary France*, ed. Constantine and Frame, p. 46.

23. Morgan and Morgan, *Travels in Revolutionary France*, ed. Constantine and Frame, p. 47.

24. Morgan and Morgan, *Travels in Revolutionary France*, ed. Constantine and Frame, p. 51.

25. Morgan and Morgan, *Travels in Revolutionary France*, ed. Constantine and Frame, p. 51.

26. Morgan and Morgan, *Travels in Revolutionary France*, ed. Constantine and Frame, p. 52.

27. Eastlake (ed.), *Dr Rigby's Letters*, p. 7.

28. Morgan and Morgan, *Travels in Revolutionary France*, ed. Constantine and Frame, p. 52.

29. Morgan and Morgan, *Travels in Revolutionary France*, ed. Constantine and Frame, p. 49.

30. Morgan and Morgan, *Travels in Revolutionary France*, ed. Constantine and Frame, p. 60.

31. Morgan and Morgan, *Travels in Revolutionary France*, ed. Constantine and Frame, p. 53.

32. Quoted in G. P. Gooch, *Germany and the French Revolution* (London and New York: Longman, Greens and Co., 1920), pp. 39–40; re-quoted in Doyle, *The Oxford History of the French* Revolution, p. 160.

33. Quoted in C. Nordmann, *Grandeur et liberté de la Suède* (1660–1792), (Paris/Louvain: Nauwelaerts, 1971), pp. 430–1; re-quoted in Doyle, *The Oxford History of the French Revolution*, p. 160.

34. Quoted in P. Dukes, 'Russia and the Eighteenth Century Revolution', *History* (1971), p. 380; re-quoted in Doyle, *The Oxford History of the French Revolution*, pp. 160–1.

35. Rogers, *Table-Talk*, p. 24.

36. William Wordsworth, *The Prelude* (London: Edward Moxon, 1850), p. 299, Book XI, lines 108–9.

37. Quoted in G. S. Veitch, *The Genesis of Parliamentary Reform* (London: Constable, 1965 edn), pp. 111–12.

38. Price had accepted an invitation to preach (see Frame, *Liberty's Apostle*, p. 217), but he recognised that he was touching 'more on political subjects than would at any other time be proper in the pulpit' (see Richard Price, *Political Writings*, ed. D. O. Thomas (Cambridge: Cambridge University Press, 1991), p. 178). He called it a Discourse and it was published as such.

39. Price, *Political Writings*, ed. Thomas, p. 177.

40. Price, *Political Writings*, ed. Thomas, pp. 178–9.

41. Price, *Political Writings*, ed. Thomas, p. 180.

42. Price, *Political Writings*, ed. Thomas, p. 195.

43. Price, *Political Writings*, ed. Thomas, pp. 195–6.

44. Morgan, *Memoirs of Richard Price*, ed. Thomas, p. 77.

45. Morgan, *Memoirs of Richard Price*, ed. Thomas, p. 78.

46. *The Gentleman's Magazine*, 59 (1789), part 2, 1121.

47. Second President of the United States (1797–1801).

48. Morgan, *Memoirs of Richard Price*, ed. Thomas, p. 79.

49. Peach (ed.), *The Correspondence of Richard Price*, vol. III, p. 282, n.4.

50. Morgan, *Memoirs of Richard Price*, ed. Thomas, p. 79.

51. See E. Beresford Chancellor, *The Annals of the Strand: Topographical and Historical* (New York: F. A. Stokes, 1912), p. 335.

52. See Veitch, *The Genesis of Parliamentary Reform*, p. 149, for details of the dinner and the toasts. Also Rémy Duthille, 'Thirteen Uncollected Letters of Richard Price', *Enlightenment and Dissent*, 27 (2011), 83–142 (97–9).

53. Morgan, *Memoirs of Richard Price*, ed. Thomas, p. 82.

54. Morgan, *Memoirs of Richard Price*, ed. Thomas, p. 83.

55. In the description and analysis of this and all the satirical cartoons referred to in this biography, I have made extensive use of the comprehensive notes made by M. Dorothy George which are provided by British Museum Satires.

56. See Tim Clayton and Sheila O'Connell, *Bonaparte and the British: prints and propaganda in the age of Napoleon* (London: British Museum, 2015), p. 25.

57. For entertaining discussions of the cartoon's ambiguity see Draper Hill, *Fashionable Contrasts: Caricatures by James Gillray* (London: Phaidon Press, 1966), p. 138, and Kenneth R. Johnston, *Unusual Suspects: Pitt's Reign of Alarm and the Lost Generation of the 1790s* (Oxford: Oxford University Press, 2013), p. 17.

58. Morgan, *Memoirs of Richard Price*, ed. Thomas, p. 86.

59. Morgan, *Memoirs of Richard Price*, ed. Thomas, p. 85.

60. Morgan, *Memoirs of Richard Price*, ed. Thomas, pp. 85–6.

Chapter 11

1. *The Gentleman's Magazine*, 61 (1791), 389–90.

2. For William's account of Price's death and funeral arrangements see Morgan, *Memoirs of Richard Price*, ed. Thomas, pp. 87–9.

3. See Frame, *Liberty's Apostle*, p. 246.

4. Morgan, *Memoirs of Richard Price*, ed. Thomas, p. 88.

5. *The Gentleman's Magazine*, 61 (1791), 390.

6. Morgan, *Memoirs of Richard Price*, ed. Thomas, pp. 89–90.

7. Joseph Priestley, *A Discourse on the Occasion of the Death of Dr Price delivered at Hackney on Sunday 1 May 1791* (London: J. Johnson, 1791), p. 9.

8. *The Gentleman's Magazine*, 61 (1791), 389–90.

9. Frame, *Liberty's Apostle*, p. 1.

10. See Carl B. Cone, *The English Jacobins: Reformers in Late 18th Century England* (New York: Charles Scribner's Sons, 1968), pp. 110–12.

11. The Dominican monastery was originally attached to the church of Sainte Jacques (no longer standing) in the Place Vendôme in Paris. The name 'Jacobin' was first applied to the club as a form of ridicule by its enemies. For an explanatory summary of French and English Jacobinism see Cone, *The English Jacobins*, pp. iii–v.

12. John Thelwall, *Rights of Nature* (1796), quoted by Cone in *The English Jacobins*, p. iii.

13. Quoted by Carl B. Cone in *Torchbearer of Freedom: The Influence of Richard Price on Eighteenth Century Thought* (Lexington: University of Kentucky Press, 1952), p. 197, and referenced as quoted in Anthony Lincoln, *Some Political and Social Ideas of English Dissent 1763–1800* (Cambridge: Cambridge University Press, 1938), p. 30, n.2.

14. Quoted by Cone in *Torchbearer of Freedom*, p. 197, and referenced as from Peter Pindar, *The Louisiad*, Canto III, in *The Works of Peter Pindar* (London: J. Walker, 1812), vol. I, p. 253.

15. Quoted by Cone in *Torchbearer of Freedom*, p. 197, and referenced as from Peter Pindar, *The Rights of Kings or Loyal Odes to Disloyal Academicians* (Dublin: William Porter, 1791), p. 24.

16. William Morgan, *A Review of Dr Price's Writings on the Subject of the Finances of this Kingdom*, p. vi.

17. Morgan, *A Review of Dr Price's Writings*, p. viii.

18. Hague, *Pitt the Younger*, p. 489.

19. Cone, *Torchbearer of Freedom*, p. 140.

20. Frame, *Liberty's Apostle*, p. 183.

21. D. O. Thomas, *The Honest Mind: The Thought and Work of Richard Price* (Oxford: Clarendon Press, 1977), p. 259.

22. See Thomas, *The Honest Mind*, pp. 236–7.

23. Morgan, *A Review of Dr Price's Writings*, pp. 63–4.

24. Morgan, *A Review of Dr Price's Writings*, p. 67.

Chapter 12

1. Rogers, *Table-Talk*, p. 111.

2. See Paula Byrne, *Perdita, The Life of Mary Robinson* (London: Harper Perennial, 2005), pp. 197–202.

3. Waugh, *A Little Learning*, p. 8. I have been unable to discover any reform club which might have issued these buttons. It could not have been the current Reform Club in Pall Mall since it did not open until 1836, twenty-four years after Horne Tooke's death.

4. See Prologue, p. xxvii.

5. Bewley, *Gentleman Radical: A Life of John Horne Tooke 1736–1812* (London and New York: Tauris Academic Studies 1998), p. 80. See also pp. 2–6 for details of Horne Tooke's early life.

6. Rogers, *Table-Talk*, p. 125.

7. Bewley, *Gentleman Radical*, p. 6.

8. Bewley, *Gentleman Radical*, p. 53.

9. William Hazlitt, *Lectures on English Poets and The Spirit of the Age* (London: J. M. Dent, Everyman edn, 1910), p. 213.

10. See Bewley, *Gentleman Radical*, pp. 60–7.

11. Price, Richard and Horne Tooke, John, *Facts: Addressed to the Landholders, Stockholders, Merchants, Farmers, Manufacturers, Tradesmen, Proprietors of every Description, and generally all The Subjects of Great Britain and Ireland* (London, 1780).

12. William Morgan, *Facts Addressed to the Serious Attention of the People of Great Britain respecting the Expence of the War and the State of the National Debt* (London: Debrett, Cadell and Davies, 1796, repr. Hardpress Classics Series), p. v.

13. See Morgan, *Memoirs of Richard Price*, ed. Thomas, pp. 40–1.

14. Thomas Jones Howell (compiler), *A Complete Collection of State Trials and Proceedings for High Treason and Other Crimes and Misdemeanors [sic] from the Year 1783 to the Present Time with Notes and Other Illustrations* (London: T. C. Hansard, Peterborough Court, Fleet Street, 1818), vol. XXV, col. 390.

15. Caroline Williams, *A Welsh Family*, pp. 139–40, describes William Morgan as being a member of the Constitutional Club. There was a Constitutional Club and a Constitutional Society (see Bewley, *Gentleman Radical*, pp. 41 and 88), but the context suggests that Caroline Williams means the Society for Promoting Constitutional Information.

16. See Veitch, *The Genesis of Parliamentary Reform*, p. 74.

17. See Veitch, *The Genesis of Parliamentary Reform*, p. 71.

18. See Veitch, *The Genesis of Parliamentary Reform*, p. 84.

19. See Veitch, *The Genesis of Parliamentary Reform*, p. 85, note 3.

20. I am very grateful to Dr Peter Davies for suggesting that I should go to Freemasons' Hall and for helping me to research Jerusalem Sols.

21. Details about the society are limited but Horne Tooke's stewardship is noted in Bro. F. W. Levander, 'The Jerusalem Sols and Some Other London Societies of the Eighteenth Century', *Ars Quatuor Coronatorum*, 25 (1912), 'Transactions of the Quatuor Cornati Lodge No. 2076 (London)'.

22. See Rev. Edward Barry, *A Sermon preached at Lambeth Church, before the Grand Modern Order of Jerusalem Sols, on their Anniversary, Thursday 17 July 1788* (London: J. Denew, 1788), a Bible-thumping address in which (p. 8) the preacher alludes to the Society's being dedicated to Solomon.

Chapter 13

1. Williams, *A Welsh Family*, p. 140.

2. Wordsworth, *The Prelude*, Book XI, lines 206–9.

3. Howell, *State Trials*, vol. XXV, col. 151.

4. Howell, *State Trials*, vol. XXV, col. 113.

5. Thomas Paine, *Rights of Man, Common Sense and Other Political Writings* (Oxford: Oxford University Press, 1995 pb. edn), p. 224.

6. Howell, *State Trials*, vol. XXII, col. 360.

7. Morgan and Morgan, *Travels in Revolutionary France*, ed. Constantine and Frame, p. 99.

8. See Morgan and Morgan, *Travels in Revolutionary France*, ed. Constantine and Frame, pp. 91–2, for a discussion of the arguments for accepting that *An Address to the Jacobine Societies* is George Morgan's work.

9. Williams, *A Welsh Family*, p. 140.

10. See Clayden, *The Early Life of Samuel Rogers*, p. 246, for an account of the dinner.

11. Rogers makes no mention of any other guest. Given that Paine suggested an alternative (and milder) toast, logic suggests that William made the toast.

12. There was more than one publication claiming to give a verbatim account of the trial of Thomas Paine. This well-known comment appears in an edition 'taken in shorthand by an eminent advocate' and 'copied from the minutes taken in court with considerable additions, corrections and alterations' (London: printed for W. Richardson, J. Parsons, C. Stalker, Mrs Harlow and William Lane, n.d.), p. 50. In the official report of the trials edited by Howells, Erskine's words are: 'From the moment that any advocate can be permitted to say, that he *will* or will *not* stand between the Crown and the subject arraigned in the court where he daily sits to practise, from that moment the liberties of England are at an end. If the advocate refuses to defend, from what *he may think* of the charge or of the defence, he assumes the character of the judge; nay he assumes it before the hour of judgment; and in proportion to his rank and reputation, puts the heavy influence of perhaps a mistaken opinion into the scale against the accused, in whose favour the benevolent principle of English law makes all presumptions, and which commands the very judge to be his counsel.' See Howell, *State Trials*, vol. XXII, col. 412.

13. The phrase used in correspondence by John Frost who accompanied Paine to France. See Alan Wharam, *The Treason Trials, 1794* (Leicester and London: Leicester University Press, 1992), p. 32.

14. For details of the Republican calendar see Clayton and O'Connell, *Bonaparte and the British*, p. 9.

15. Hague, *Pitt the Younger*, p. 329.

16. George Cadogan Morgan, despite his fiery republicanism, suggested retiring the deposed king to a safe enclosure. See Morgan and Morgan, *Travels in Revolutionary France*, ed. Constantine and Frame, pp. 95 and 116.

17. Veitch, *The Genesis of Parliamentary Reform*, p. 192.

18. David Vincent (ed.), *Testaments of Radicalism: Memoirs of Working Class Politicians 1790–1885* (London: Europa Publications Ltd, 1977), p. 45.

19. Francis Place, *The Autobiography of Francis Place*, ed. Mary Thale (Cambridge: Cambridge University Press, 1972), p. 139.

20. See Wharam, *The Treason Trials*, pp. 48–67.

21. Howell, *State Trials*, vol. XXIV, cols 745–8.
22. Howell, *State Trials*, vol. XXV, col. 4.
23. The Crown's witness, William Sharpe, when cross-examined by Horne Tooke, confirmed this:
    'Horne Tooke: do you remember what was the subject of that letter?
    Sharpe: Yes, I do; it alluded to the red book, the Court Calendar – there were to be some extracts made of the sinecure places and pensions that Mr Pitt and his family received from the public; and Mr Joyce called upon me the day upon which Mr Hardy was taken up, and desired you to be ready at Spitalfields, on Thursday next.'
    See Howell, *State Trials*, vol. XXV, cols 248–9.
24. As well as Horne Tooke and Thomas Hardy, the twelve apostles included: John Thelwall (1764–1834), writer and lecturer; Rev. Jeremiah Joyce (1763–1816), Unitarian minister and tutor to Lord Stanhope's sons; John Augustus Bonney (1763–1813), lawyer; John Richter (1769?–1830), gentleman; Steward Kydd (1759–1811), barrister-at-law of the Middle Temple; Thomas Holcroft (1745–1809), playwright; Matthew Moore, gentleman; Thomas Wardle, gentleman; Richard Hodgson, Westminster hatter; John Baxter, labourer. See Veitch, *The Genesis of Parliamentary Reform*, pp. 307–9.
25. See Bewley, *Gentleman Radical*, p. 154.
26. Williams, *A Welsh Family*, p. 139.
27. The title 'King of France' was used until 1 January 1801. See Chris Cook and John Stevenson, *British Historical Facts 1760–1830* (London: Macmillan, 1980), p. 2.
28. Joyce is the more usual spelling.
29. John Debrett was later to become editor of *Debrett's Peerage*.
30. See A. V. Beedell and A. D. Harvey (eds), *The Prison Diary (16 May–22 November 1794) of John Horne Tooke* (Leeds: Leeds Philosophical and Literary Society Ltd, 1995), p. 9.
31. Beedell and Harvey (eds), *The Prison Diary of John Horne Tooke*, p. 56.
32. See Beedell and Harvey (eds), *The Prison Diary of John Horne Tooke*, pp. 33–5.
33. A hard tumour, possibly an early stage of cancer
34. Beedell and Harvey, *The Prison Diary of John Horne Tooke*, pp. 46–7.
35. Alexander Stephens, *Memoirs of John Horne Tooke*, vol. II (London: J. Johnson and Co, 1813), pp. 420–3, and Bewley, *Gentleman Radical*, p. 158.
36. Williams, *A Welsh Family*, p. 38. A handwritten note in author's copy corrects 'Clive' to 'Cline' (which, by deduction, identifies surgeon).
37. Williams, *A Welsh Family*, p. 136.
38. See 'Thomas Hardy, *Memoir* (London: James Ridgway, 1832)', in David Vincent (ed.), *Testaments of Radicalism: Memoirs of Working Class Politicians 1790–1885* (London: Europa Publications Ltd, 1977), pp. 64–5.
39. Christina and David Bailey, *Gentleman Radical*, p. 168.

Chapter 14

1. Howell, *State Trials*, vol. XXV, col. 8.
2. *OED*, 'A loge is a box, the term more usually used for a box at the theatre'.
3. Rogers, *Table-Talk*, p. 20.

4. Cecilia Lucy Brightwell, *Memorials of the Life of Amelia Opie, selected and arranged from her Letters, Diaries and Other Manuscripts* (Norwich: Fletcher and Alexander, 1854), p. 47.

5. Brightwell, *Memorials of Amelia Opie*, p. 46.

6. There was first the grand jury of twenty-one men whose function was to decide if there was a 'true bill of indictment' against the accused. Once the true bill had been established, the trial took place. For this a jury of twelve was sworn in. See Wharam, *The Treason Trials*, pp. 132 and 146–7.

7. See Wharam, *The Treason Trials*, pp. 143–6.

8. See Wharam, *The Treason Trials*, pp. 155–6.

9. For a discussion of the accuracy of the shorthand notes and the transcripts see Susan Mitchell Sommers, *The Siblys of London, A Family on the Esoteric Fringes of Georgian England* (Oxford: Oxford University Press, 2018), pp. 18–21.

10. See Howell, *State Trials*, vol. XXIV, col. 201. The Attorney General referred to a statute passed in the twenty-fifth year of the reign of Edward III: 'by that statute it is declared to be high treason to compass or imagine the death of the king ... Not only acts of immediate and direct attempt against the king's life are overt acts of compassing his death, but that all the remoter steps taken with a view to assist to bring about the actual attempt, are equally overt acts of this species of treason'.

11. Howell, *State Trials*, vol. XXIV, col. 681: Erskine, 'It was fabricated by spies who support the prosecution.'

12. Howell, *State Trials*, vol. XXIV, cols 682–3.

13. Wharam, *The Treason Trials*, p. 171.

14. Wharam, *The Treason Trials*, p. 173.

15. Manoah Sibly, *The Genuine Trial of Thomas Hardy for High Treason at the Sessions House in the Old Bailey from October 28 to November 5, 1794, Vol. II* (London: J. S. Jordan, 1795), p. 590.

16. Vincent, *Testaments of Radicalism*, p. 72.

17. Samuel Rogers, *Human Life, A Poem* (London: John Murray, 1819), p. 49.

18. Howell, *State Trials*, vol. XXV, col. 8.

19. *The Gentleman's Magazine*, 64 (1794), part 2, 1050.

20. Howell, *State Trials*, vol. XXV, col. 381.

21. See chapter 12, 'A Radical Friend'.

22. Howell, *State Trials*, vol. XXV, cols 385–6.

23. Howell, *State Trials*, vol. XXV, cols 330–1.

24. William Hazlitt, *The Spirit of the Age*, in *The Works of William Hazlitt*, The World's Classics (London: Grant Richards, 1904), p. 68.

25. Williams, *A Welsh Family*, p. 136.

Chapter 15

1. Morgan, *Facts Addressed to the Serious Attention of the People of Great Britain respecting the Expence of the War and the State of the National Debt*, p. iv.

2. Minister 'at War', not 'of War'.

3. Johnston, *Unusual Suspects*, p. 333, note iv.

4. See Wharam, *The Treason Trials*, pp. 246–7.

5. Bewley, *Gentleman Radical*, p. 268.

6. P. W. Clayden, *Rogers and His Contemporaries*, vol 1 (London: Smith, Elder and Co., 1889), p. 83.

7. Stephens, *Memoirs of John Horne Tooke*, p. 449.

8. Letter of 26 December 1812 to John Price in the Glamorgan Archives, Merthyr Mawr Estate Collection, DM/PR/52/16.

9. See Veitch, *The Genesis of Parliamentary Reform*, pp. 324–7.

10. Williams, *A Welsh Family*, p. 139.

11. See Jenny Uglow, *In These Times: Living in Britain Through Napoleon's Wars 1793–1815* (London: Faber and Faber, 2015), p. 141.

12. Francis Place gives an eyewitness account of the event in his autobiography. See Place, *Autobiography of Francis Place*, ed. Thale, pp. 146–7.

13. Morgan, *Facts Addressed to the Serious Attention of the People of Great Britain respecting the Expence of the War and the State of the National Debt*, pp. iii–iv.

14. Morgan, *Facts . . . respecting the Expence of the War*, pp. 1–3.

15. Morgan, *Facts . . . respecting the Expence of the War*, p. 10.

16. Morgan, *Facts . . . respecting the Expence of the War*, pp. 40–4.

17. Society for Equitable Assurances: William Morgan, *The Deed of Settlement of the Society for Equitable Assurances on Lives and Survivorships, as the same is inrolled in His Majesty's Court of King's Bench at Westminster in the year 1765; with the bye-laws and orders: To which are appended reports by the court of directors, and nine addresses by William Morgan Esq., F.R.S., late Actuary of the Society* (London: printed by Richard Taylor, 1833), pp. 198–9 (abbreviated to *Nine Addresses*).

18. Elderton, 'William Morgan', in *Transactions of the Faculty of Actuaries*, 14 (1931–4), 4.

19. The letter is in the author's collection.

20. Bagwell, *The Transport Revolution*, pp. 40–1.

21. The beach was probably at Southerndown, much loved by William, and with large stretches of sand.

22. Williams, *A Welsh Family*, p. 71.

23. Letter of 29 December to Catherine Price, Glamorgan Archives, Merthyr Mawr Estate Collection DM/PR/52/1.

Chapter 16

1. William Morgan, *An Appeal to the People of Great Britain, on the Present Alarming State of the Public Finances, and of Public Credit* (London: J. Debrett, 1797, repr. Gale ECCO), p. 80.

2. See Uglow, *In These Times*, pp. 163–9.

3. Although the local women would have been present as spectators, the delightful story of their role in deceiving the invaders is probably more legend than accurate fact. See J. E. Thomas, *Britain's Last Invasion, Fishguard 1797* (Stroud: Tempus Publications Ltd, 2007), pp. 149–58.

4. See Hague, *William Pitt*, pp. 397–8.

5. See Hague, *William Pitt*, p. 398.

6. Morgan, *An Appeal to the People of Great Britain*, p. 64.

7. See Clayton and O'Connell, *Bonaparte and the British*, p 63, for further analysis of the cartoon.

8. The full title is *An Appeal to the People of Great Britain on the present alarming state of the Public Finances and of Public Credit*.

9. Full title, *Facts Addressed to the Serious Attention of the People of Great Britain respecting the Expence [sic] of the War and the State of the National Debt*.

10. Morgan, *An Appeal to the People of Great Britain* (second edn), p. 10.

11. Santo Domingo is the capital and chief seaport of the Dominican Republic. Strategically important, it was fought over and changed hands several times during the late eighteenth and early nineteenth centuries.

12. Morgan, *An Appeal to the People of Great Britain* (second edn), pp. 12–13.

13. Morgan, *An Appeal to the People of Great Britain* (second edn), p. 13.

14. Morgan, *An Appeal to the People of Great Britain* (second edn), p. 60.

15. Morgan, *An Appeal to the People of Great Britain* (second edn), p. 34.

16. Morgan, *An Appeal to the People of Great Britain* (second edn), p. 80.

17. Morgan, *An Appeal to the People of Great Britain* (second edn), p. 78.

18. See the speech made by William Scott, brother-in-law to the Earl of Oxford, at the Hampshire meeting on 19 April 1797, in Michael T. Davis (ed.), *London Corresponding Society, 1792–1799, Vol. 4* (London: Pickering and Chatto, 2002), p. 221.

19. See Johnston, *Unusual Suspects*, pp. 8 and 190, note ii.

20. Williams, *A Welsh Family*, p. 122.

Chapter 17

1. William Morgan, *A Comparative View of the Public Finances from the Beginning to the Close of the Late Administration* (London: J. Debrett, 1801, repr. Hardpress Publishing), p. 75.

2. Williams, *A Welsh Family*, p. 125.

3. L. G. Johnson, *General T. Perronet Thompson 1783–1869, His Military, Literary and Political Campaigns* (London: George, Allen and Unwin Ltd, 1957), p. 14.

4. Williams, *A Welsh Family*, pp. 125–6.

5. Williams, *A Welsh Family*, p. 126.

6. William Morgan, 'Electrical Experiments Made in Order to Ascertain the Non-Conducting Power of a Perfect Vacuum', in *Philosophical Transactions of the Royal Society*, 75 (1785), 276.

7. Lyndall Gordon, *Mary Wollstonecraft: A New Genus* (London: Little, Brown and Co., 2005), p. 43.

8. 'Account of the late Mr George Cadogan Morgan [Dec. 1798]', in *The Monthly Magazine and British Register*, no. XXXIX, vol. VI (London: printed for R. Phillips, December 1798), 478–9.

9. Williams, *A Welsh Family*, p. 129.
10. 'Account of the late Mr George Cadogan Morgan [Dec. 1798]', in *The Monthly Magazine and British Register*, 479.
11. Morgan and Morgan, *Travels in Revolutionary France*, ed. Constantine and Frame, p. 144.
12. See Johnston, *Unusual Suspects*, p. xiv and Appendix 1, pp. 329–30.
13. Clayden, *The Early Life of Samuel Rogers*, p. 290.
14. Williams, *A Welsh Family*, p. 136. Williams states that William 'was on particularly friendly terms with Watson, the Bishop of Llandaff'. I find it hard to believe that he remained on such friendly terms after Watson revised his views.
15. Johnston, *Unusual Suspects*, p. 190.
16. Clayden, *Rogers and His Contemporaries*, vol. ii, p. 232.
17. Williams, *A Welsh Family*, p. 122.
18. Also called *The Weekly Examiner*.
19. The *Anti-Jacobin* arrived at this figure by multiplying the regular weekly sale of 2,500 by seven (considered to be the average size of a family) thus arriving at 17,500, then adding 32,500 based on the claim that many readers lent their copies to their poorer neighbours. See Wendy Hinde, *George Canning* (London: Collins, 1975), p. 65.
20. *The Anti-Jacobin*, no. 3 (November 1797), 104.
21. Morgan, *A Comparative View of the Public Finances* (second edn), p. iv.
22. Morgan, *A Comparative View of the Public Finances* (second edn), p. 2.
23. Morgan, *A Comparative View of the Public Finances* (second edn), p. 39.
24. Morgan, *A Comparative View of the Public Finances* (second edn), p. 75.
25. See Ogborn, *Equitable Assurances*, p. 128.
26. Ogborn, *Equitable Assurances*, p. 129.

Chapter 18

1. See Paula Byrne, *The Real Jane Austen: A Life in Small Things* (London: William Collins, 2014 pb. edn), p. 14.
2. Samuel Boddington, *Treachery and Adultery, £10,000 Damages! Trial of Benjamin Boddington, Esq. for adultery with Mrs Boddington, his cousin's wife before the Sheriff of London, on Friday the 8th of September, 1797* (London: printed at No. 8, White-Hart-Yard, Strand, 1797, repr. Gale ECCO), p. 4.
3. Morgan and Morgan, *Travels in Revolutionary France*, ed. Constantine and Frame, p. 144.
4. Quoted from an unpublished family memoir attributed to Anne Ashburner, granddaughter of George Cadogan Morgan, in the Ashburner Papers, Sedgwick Collection, Proctor Museum and Archives, Stockbridge Library, Stockbridge, MA.
5. See Morgan and Morgan, *Travels in Revolutionary France*, ed. Constantine and Frame, pp. 156–8, for a full account of the journey.
6. Morgan and Morgan, *Travels in Revolutionary France*, ed. Constantine and Frame, p. 185.

Chapter 19

1. Letter of 26 December 1815 to John Price, Glamorgan Archives, Merthyr Mawr Estate Collection DM/PR/52/24.
2. Morgan, *Nine Addresses*, p. 208.
3. Ogborn, *Equitable Assurances*, p. 121.
4. Morgan, *Nine Addresses*, p. 223.
5. Morgan, *Nine Addresses*, p. 222.
6. *The Times*, 2 May 1933.
7. Letter of 5 March 1807 to John Price, Glamorgan Archives, Merthyr Mawr Estate Collection, DM/PR/52/8.
8. Morgan, *Nine Addresses*, p. 233.
9. Minutes of the General Court for 7 December 1809.
10. Letter of 2 January 1810 to John Price, Glamorgan Archives, Merthyr Mawr Estate Collection, DM/PR/52/12.
11. Minutes of the General Court for 7 December 1786, pp. 204–6 (where the minutes have page numbers these are included, but not all are numbered).
12. Minutes of the General Court for 6 December 1792.
13. Minutes of the General Court for 1 September 1791.
14. Ogborn, *Equitable Assurances*, pp. 152–3.
15. Ogborn, *Equitable Assurances*, pp. 153–4.
16. Letter of 3 October 1815 to John Price, Glamorgan Archives, Merthyr Mawr Estate Collection, DM/PR/52/23.
17. John Wade, *History of the Middle and Working Classes; with a Popular Exposition of the Economical and Political Principles which have influenced the Past and Present Condition of the Industrious Orders* (London: Effingham Wilson, 1833), p. 559 (note).
18. Morgan, *A View of the Rise and Progress of the Equitable Society*, pp. 41–2 and 45–6. See also Turnbull, *The History of British Actuarial Thought*, p. 7.
19. See Turnbull, *The History of British Actuarial Thought*, pp. 72–80, for a detailed discussion of the shortcomings of William's advice.
20. Morgan, *A View of the Rise and Progress of the Equitable Society*, p. 5.

Chapter 20

1. John Feltham, *A Guide to All the Watering Places and Sea-bathing Places; with a Description of the Lakes; a Sketch of a Tour in Wales; and Itineraries, illustrated with maps and views* (London: printed for Richard Phillips, 1806), p. 507.
2. Feltham, *A Guide to All the Watering Places and Sea-bathing Places*, p. 344.
3. Letter of 26 September 1804, in author's private collection.
4. Feltham, *A Guide to All the Watering Places and Sea-bathing Places*, p. 328.
5. Letter of 26 September 1804, in the author's private collection.
6. Morgan and Morgan, *Travels in Revolutionary France*, ed. Constantine and Frame, p. 145.
7. Letter of 22 July 1811, in author's private collection.

8. I am grateful to John Morgan, another descendant of William Morgan, for allowing me to quote from this letter of 28 November 1812 in his private collection.

9. London Metropolitan Archive, 'Faculty for removing the corpse of Mrs Sarah Travers from the Churchyard', catalogue reference DRO/020/B/02/002.

10. I am grateful to Paul Frame for allowing me to quote from this inscription.

11. Undated letters in the author's private collection.

12. *OED*, 'A close fitting jacket or bodice commonly worn by women and children early in the nineteenth century'.

13. Letter of 28 July 1831 quotes from Byron's 'The Dream', written about the poet's unrequited love for his kinswoman Mary Ann Chaworth. See Benita Eisler, *Byron: Child of Passion, Fool of Fame* (London: Hamish Hamilton, 1999), pp. 67 and 69–71.

Chapter 21

1. Quoted by Samuel Rogers in *Table-Talk*, p. 132.

2. Letter of 9 April 1814 to John Price, Glamorgan Archives, Merthyr Mawr Estate collection, DM/PR/52/29.

3. See John Wade, *British History Chronologically Arranged; Comprehending a Classified Analysis of Events and Occurrences in Church and State; and of the Constitutional, Political, Commercial, Intellectual and Social Progress of the United Kingdom from the First Invasion by the Romans to the Accession of Queen Victoria* (London: Effingham Wilson, Royal Exchange, 1839), p. 714.

4. Wade, *British History Chronologically Arranged*, p. 723.

5. Morgan, *Memoirs of the Life of Richard Price*, ed. Thomas, pp. iii–iv.

6. See chapter 5, 'At War'.

7. Morgan, *Memoirs of the Life of Richard Price*, ed. Thomas, p. 17.

8. Morgan, *Memoirs of the Life of Richard Price*, ed. Thomas, p. 17.

9. Morgan, *Memoirs of the Life of Richard Price*, ed. Thomas, p. 40.

10. Morgan, *Memoirs of the Life of Richard Price*, ed. Thomas, p. 41.

11. Morgan, *Memoirs of the Life of Richard Price*, ed. Thomas, p. 24.

12. Morgan, *Memoirs of the Life of Richard Price*, ed. Thomas, p. 24.

13. Morgan, *Memoirs of the Life of Richard Price*, ed. Thomas, p. 70.

14. Morgan, *Memoirs of the Life of Richard Price*, ed. Thomas, p. iii.

15. Morgan, William, *Sermons on Various Subjects, by the late Dr. Richard Price D.D., F.R.S.* (London: Longman, Hurst, Rees, Orme and Brown, 1816).

16. Morgan, *Memoirs of the Life of Richard Price*, ed. Thomas, p. 54.

17. Waugh, *A Little Learning*, p. 8.

18. Morgan, *Memoirs of the Life of Richard Price*, ed. Thomas, p. 94.

19. Ogborn, *Equitable Assurances*, p. 155.

20. Minutes of the General Court for 5 December 1816.

21. Minutes of the General Court for 5 June 1817.

22. Letter of 8 May 1817 to John Price, Glamorgan Archives, Merthyr Mawr Estate Collection, DM/PR/52/31.

23. Waugh, *A Little Learning*, p. 8.

24. Waugh, *A Little Learning*, p. 8.

25. Policy No. 20439, WCoD, 26 November 1802.

26. I am indebted to Robin Simon for his help in reading the signs in the portrait.

27. Letter of 20 March 1817, Glamorgan Archives, Merthyr Mawr Estate Collection, DM/PR/52/29.

28. Williams, *A Welsh Family*, pp. 136–7.

29. Its full name is Scottish Widows' Fund and Equitable Assurance Society. See Ogborn, *Equitable Assurances*, p. 198.

30. House of Commons, *Report from the Select Committee on the Poor Laws: with the minutes of evidence taken before the committee: and an appendix* (London, July 1817), pp. 135–6. Available at *https://babel.hathitrust.org/cgi/pt?id=nyp.33433000256689 &view=1up&seq=142*. Accessed 2018.

31. Letter of 6 August 1817 to Jane Price, Glamorgan Archives, Merthyr Mawr Estate Collection, DM/PR/52/34.

32. Letter of 1818 (no month given) to Jane Price, Glamorgan Archives, Merthyr Mawr Estate Collection, DM/PR/52/36.

33. Morgan, *Nine Addresses*, p. 266.

34. Minutes of the General Court for 2 December 1819.

35. Caroline Williams makes this assertion in *A Welsh Family*, p. 138.

Chapter 22

1. Morgan, *A View of the Rise and Progress of the Equitable Society*, p. 55.

2. Sir Charles Morgan and William Morgan were not related.

3. Ogborn, *Equitable Assurances*, p. 175.

4. Letter of 2 August 1824, in author's private collection.

5. Ogborn, *Equitable Assurances*, p. 183.

6. Morgan, *Nine Addresses*, pp. 284–5.

7. See Ogborn, *Equitable Assurances*, pp. 176–88, for the names of some of the activists.

8. Letter held at the British Library, Add 37182 f 173.

9. Charles Babbage, *Reflections on the Decline of Science in England, and on Some of its Causes* (London: B. Fellowes and J. Booth, 1830).

10. Charles Babbage, *A Word to the Wise* (1833) (London: J. Murray, 1856).

11. Charles Babbage, *A Comparative View of the Various Institutions for the Assurance of Lives* (London: printed for J. Mawman, 1826, repr. Gale MOML), p. x.

12. Letter held in British Library, Add 37184 f 24, 312, 362.

13. Morgan, *A View of the Rise and Progress of the Equitable Society*.

14. Morgan, *A View of the Rise and Progress of the Equitable*, pp. 44–7.

15. Morgan, *A View of the Rise and Progress of the Equitable*, p. 55.

16. Morgan, *A View of the Rise and Progress of the Equitable*, p. 59.

17. See Morgan, *The Rise and Progress of the Equitable Society*, note on p. 51.

18. Ogborn, *Equitable Assurances*, p. 192.

Chapter 24

1. *The Times*, 2 May 1933.
2. See Clayden, *Rogers and His Contemporaries*, vol. ii, p. 132.
3. Morgan, *Nine Addresses*, p. 296.
4. Letter of 21 April 1831 from William to Mrs Huddy, in the author's private collection.
5. Morgan, *Nine Addresses*, pp. 297–9.
6. See Christopher Hibbert, *George IV: Regent and King 1811–1830* (London: Allen Lane, 1973), p. 342.
7. Quoted by Ann Thwaite in *Glimpses of the Wonderful* (London: Faber and Faber, 2002), p. 340, note to p. 13.
8. Tennyson, *In Memoriam*, section LV, in *The Works of Tennyson* (London: Strahan and Co, 1871), p. 80.
9. Tennyson, *In Memoriam*, section LVI, in *The Works of Tennyson*, p. 82.
10. Letter of 24 October 1885 from Philip Henry Gosse in Sandhurst, Torquay, to his cousin Susan Gosse (1814–97) in Clifton, Bristol.
11. See Ann Thwaite, *Edmund Gosse: A Literary Landscape 1849–1928* (London: Secker and Warburg, 1984), p. 7.
12. Ogborn, *Equitable Assurances*, p. 205.

Chapter 25

1. Letter quoted in Clayden, *Rogers and His Contemporaries*, vol. ii, p. 352.
2. Clayden, *Rogers and His Contemporaries*, vol. ii, p. 353.
3. William and Arthur between them clocked up ninety-five years as actuary to the Equitable.
4. Obituary in *The London Medical Gazette*, vol. V (London: Longman, Brown, Green and Longmans, 1847), p. 778.
5. Pamela Bright, *Dr Richard Bright (1789–1858)* (London: Bodley Head, 1983), p. 237.
6. Obituary in *The London Medical Gazette*, vol. V, p. 778.
7. Obituary in *The London Medical Gazette*, vol. V, p. 779.
8. Waugh, *One Man's Road*, p. 12.
9. Hackney Archives, Scrapbook of Florence Bagust, vol. 14, p. 161.
10. Hackney Archives, Scrapbook of Florence Bagust, vol. 14, p. 161.

Epilogue

1. At-Bristol has now been renamed as We the Curious.

BIBLIOGRAPHY

Primary sources

William Morgan's papers to the Royal Society
'Electrical Experiments Made in Order to Ascertain the Non-Conducting Power of a Perfect Vacuum &c', Communicated by the Rev. Richard Price, LL.D.F.R.S., *Philosophical Transactions of the Royal Society*, 75 (London: Royal Society, 1785), 272–8.

'On the Probabilities of Survivorship between Two Persons of Any Given Ages, and the Method of Determining the Values of Reversions Depending on Those Survivorships', Communicated by the Rev. Richard Price, D.D.F.R.S., *Philosophical Transactions of the Royal Society*, 78 (London: Royal Society, 1788), 331–49.

'On the Method of Determining, from the Real Probabilities of Life, the Value of a Contingent Reversion in Which Three Lives are Involved in the Survivorship', Communicated by the Rev. Richard Price, D.D.F.R.S., *Philosophical Transactions of the Royal Society*, 79 (London: Royal Society, 1789), 40–54.

'On the Method of Determining, from the Real Probabilities of Life, the Value of Contingent Reversions in Which Three Lives are Involved in the Survivorship', *Philosophical Transactions of the Royal Society*, 84 (London: Royal Society, 1791), 223–61.

'On the Method of Determining, from the Real Probabilities of Life, the Value of Contingent Reversions in Which Three Lives are Involved in the Survivorship', *Philosophical Transactions of the Royal Society*, 81 (London: Royal Society, 1794), 246–77.

'On the Method of Determining, from the Real Probabilities of Life, the Value of Contingent Reversions in Which Three Lives are Involved in the Survivorship', *Philosophical Transactions of the Royal Society*, 90 (London: Royal Society, 1800), 22–45.

William Morgan's books and pamphlets

Several are in modern reproductions, abbreviated as follows:

Gale Eighteenth Century Collections Online Print Editions – Gale ECCO

Gale The Making of Modern Law Print Editions – Gale MOML

Hardpress Classics Series – Hardpress

The doctrine of Annuities and Assurances on Lives and Survivorships, stated and explained. To which is added an Introduction, addressed to the Society. Also An Essay on the Present State of Population in England and Wales, by the Reverend Dr Price (London: T. Cadell, 1779, in Gale ECCO).

An Examination of Dr Crawford's Theory of Heat and Combustion (London: T. Cadell, 1781, in Gale ECCO).

A Review of Dr Price's Writings, on the subject of the Finances of this Kingdom: to which are added the Three Plans communicated by him to Mr Pitt in the year 1786, for Redeeming the National Debt: and also an Enquiry into the Real State of the Public Income and Expenditure, from the Establishment of the Consolidated Fund to the year 1791 (London: G. Stafford, 1792, in Gale ECCO). Second edition, with *A Supplement stating the Amount of the Debt in 1795* (1796).

Facts addressed to the Serious Attention of the People of Great Britain respecting the Expence [sic] *of the War and the State of the National Debt* (London: J. Debrett, T. Cadell and W. Davies, 1796, in Hardpress). Four editions were published in 1796.

Additional Facts, addressed to the Serious Attention of the People of Great Britain respecting the Expences [sic] *of the War and the State of the National Debt* (London: J. Debrett, T. Cadell and W. Davies, 1796). Four editions, second in Gale ECCO.

An Appeal to the People of Great Britain, on the Present Alarming State of the Public Finances, and of Public Credit (London J. Debrett, 1797). Four editions, second in Gale ECCO.

A Comparative View of the Public Finances from the beginning to the close of the Late Administration (London: J. Debrett, 1801). Three editions, third in Hardpress.

Memoirs of the Life of the Rev Richard Price (London: R. Hunter and R. Rees, 1815; D. O. Thomas (ed.), in *Enlightenment and Dissent*, 22 (2003)).

Sermons on Various Subject, by the late Dr. Richard Price D. D., F. R. S. (London: Longman, Hurst, Rees, Orme and Brown, 1816).

The Principles and Doctrines of Assurances; Annuities on Lives, and Contingent Reversions, Stated and Explained (London: Longman, Hurst, Rees, Orme and Brown, 1821; now printed and digitised by General Books).

A View of the Rise and Progress of the Equitable Society and of the Causes which have contributed to its Success with Remarks on some of the Late Misrepresentations respecting the Rules and Practice of the Society (London: Longman, Rees, Orme, Brown and Green, 1829). Two editions, second in Gale MOML.

Nine Addresses to Members of the Equitable Society included in *The Deed of Settlement of the Society for Equitable Assurances on Lives and Survivorships, as the same is inrolled in His Majesty's Court of King's Bench at Westminster in the year 1765; with the bye-laws and orders: To which are appended reports by the court of directors, and nine addresses by William Morgan Esq., F.R.S., late Actuary of the Society* (London: Printed by Richard Taylor, 1833).

Secondary sources

Ackroyd, Peter, *London: The Biography* (London: Chatto and Windus, 2000).

Ackroyd, Peter, *The History of England, Vol. 1V: Revolution* (London: Macmillan, 2016).

Albinson, Cassandra, Funnell, Peter, Pointon, Marcia, and Peltz, Lucy, *Thomas Lawrence: Regency Power and Brilliance* (London: National Portrait Gallery, 2010).

Allardyce, Alex, *The Village that Changed the World: A History of Newington Green N 16* (London: Newington Green Action Group, 2008).

Allen, Rick, *The Moving Pageant: A Literary Sourcebook on London Street-life, 1700–1914* (London and New York: Routledge, 1998).

Aspland, R. (ed.), *The Christian Reformer, or, Unitarian Magazine and Review* (London: Edward T. Whitfield, 1856), vol. XII, no. CXXXVI.

Babbage, Charles, *A Comparative View of the Various Institutions for the Assurance of Lives* (London: J. Mawman, 1826) (Gale MOML).

Babbage, Charles, *Science and Reform: Selected Works of Charles Babbage*, ed. and intr. Anthony Hyman (Cambridge: Cambridge University Press, 1989).

Bagwell, Philip S., and Lyth, Peter, *Transport in Britain: From Canal Lock to Gridlock* (London: Hambledon, 2002).

Bagwell, Philip S., *The Transport Revolution from 1770* (London: B. T. Batsford Ltd, 1974).

Baker, Kenneth, *George III: A Life in Caricature* (London: Thames and Hudson, 2007).

Barry, Rev. Edward, *A Sermon preached at Lambeth Church, before the Royal Grand Modern Order of Jerusalem Sols, on their Anniversary, Thursday 17th July 1788* (London: J. Denew, 1788, in Gale ECCO).

Bayne-Powell, Rosamund, *Eighteenth Century London Life* (London: John Murray, 1937)

Bayne-Powell, Rosamund, *Travellers in Eighteenth Century England* (London: John Murray, 1951).

Beedell, A. V., and Harvey, A. D. (eds), *The Prison Diaries of John Horne Tooke* (Leeds: Leeds Philosophical and Literary Society Ltd, 1995).

Bellhouse, David R., *Leases for Lives: Life contingent contracts and the emergence of actuarial science in eighteenth-century England* (Cambridge: Cambridge University Press, 2017).

Bewley, Christina and David, *Gentleman Radical: A Life of John Horne Tooke 1736–1812* (London and New York: Tauris Academic Studies, 1998).

Boddington, Samuel, *Treachery and Adultery, £10,000 Damages! Trial of Benjamin Boddington, Esq. for adultery with Mrs Boddington, his cousin's wife before the Sheriff of London, on Friday the 8th of September, 1797* (London: printed at No. 8, White-Hart-Yard, Strand, 1797; in Gale ECCO).

Braithwaite, Helen, *Romanticism, Publishing and Dissent: Joseph Johnson and the Cause of Liberty* (Basingstoke: Palgrave Macmillan, 2003).

Bright, Pamela, *Dr Richard Bright 1789–1858* (London, Sydney, Toronto: Bodley Head, 1983).

Brightwell, Cecilia Lucy, *Memorials of the Life of Amelia Opie, selected and arranged from her Letters, Diaries and Other Manuscripts* (Norwich: Fletcher and Alexander, 1854).

Byrne, Paula, *Perdita, The Life of Mary Robinson* (London: HarperPerennial, 2005).

Byrne, Paula, *The Real Jane Austen: A Life in Small Things* (London: William Collins, pb. edn, 2014)

Bryson, Bill (ed.), *Seeing Further: The Story of Science and the Royal Society* (London: Harper Press, 2010).

Chancellor, E. Beresford, *The Annals of the Strand: Topographical and Historical* (New York: F. A. Stokes, 1912).

Clark, Peter, *British Clubs and Societies 1580–1795: The origins of an associational world* (Oxford: Clarendon Press, 2000).

Clark, Peter, *Sociability and Urbanity: Clubs and societies in the eighteenth century city* (Leicester: Victorian Studies Centre, University of Leicester, 1986).

Clayden, P. W., *The Early Life of Samuel Rogers* (London: Smith, Elder and Co., 1887).

Clayden, P. W., *Rogers and His Contemporaries* (London: Smith, Elder and Co., 1889).

Clayton, Tim, and O'Connell, Sheila, *Bonaparte and the British: prints and propaganda in the age of Napoleon* (London: British Museum, 2015).

Cobban, Alfred, *The Debate on the French Revolution* (London: A. and C. Black, 1960).

Cohen, I. Bernard, *Benjamin Franklin's Science* (Cambridge, MA, and London: Harvard University Press, 1990)

Cone, Carl B., *The English Jacobins: Reformers in Late 18th Century England* (New York: Charles Scribner's Sons, 1968).

Cone, Carl B., *Torchbearer of Freedom: The Influence of Richard Price on Eighteenth Century Thought* (Lexington: University of Kentucky Press, 1952).

Cook, Chris, and Stevenson, John, *British Historical Facts 1760–1830* (London: Macmillan, 1980).

Copeman, W. S. C., *The Worshipful Society of Apothecaries of London: A History 1617–1967* (Oxford: Pergamon Press 1967).

Davies, Iolo, *A Certaine Schoole, A History of the Grammar School at Cowbridge Glamorgan* (Cowbridge: D. Brown and Sons Ltd, 1967).

Davis, Michael T. (ed.), *London Corresponding Society, 1792–1799, Vol. 4* (London: Pickering and Chatto, 2002).

De Castro, J. Paul, *The Gordon Riots* (Oxford: Oxford University Press, 1926).

De Saussure, César, *A Foreign View of England in the Reigns of George I and George II*, trans. and ed. Madame van Muyden (London: J. Murray, 1902).

Derry, John W., *The Radical Tradition: Tom Paine to Lloyd George* (London: Macmillan, 1967).

Dickens, Charles, *Barnaby Rudge* (London: Everyman's Library, 2005 edn).

Doyle, William, *The Oxford History of the French Revolution* (Oxford and New York: Oxford University Press, 1989).

Eisler, Benita, *Byron: Child of Passion, Fool of Fame* (London: Hamish Hamilton, 1999).

Elderton, William Palin, 'William Morgan, F.R.S. 1750–1833', in *Transactions of the Faculty of Actuaries* (1931–4), vol. 14.

Elderton, William Palin, 'Some Family Connections of William Morgan (1750–1833) F.R.S.', in *The Genealogists' Magazine*, 12/10 (June 1957).

Evans, Eric J., *The Forging of the Modern State: Early Industrial Britain 1783–1870* (third edn, Harlow: Longman, 2001).

Feltham, John, *A Guide to All the Watering Places and Sea-bathing Places; with a Description of the Lakes; a Sketch of a Tour in Wales; and Itineraries, illustrated with maps and views* (London: printed for Richard Phillips, 1806).

Ferguson, Niall, *The World's Banker: The History of the House of Rothschild* (London: Weidenfeld & Nicolson, 1998).

Frame, Paul, *Liberty's Apostle: Richard Price, His Life and Times* (Cardiff: University of Wales Press, 2015).

Franklin, Benjamin, *Autobiography and Other Writings* (Oxford: Oxford University Press, Oxford World Classics edn, 1998).

Franklin, Benjamin, *The Papers 1706–1790* (New Haven: Yale University Press, 1959).

Gardiner, Rena, *The Story of St Bartholomew the Great* (London: Parish of St Bartholomew the Great, 2005).

Garlick, Kenneth, *Sir Thomas Lawrence: A Complete Catalogue of the Oil Paintings* (Oxford: Phaidon, 1989).

George, M. Dorothy, *London Life in the Eighteenth Century* (London: Peregrine Books pb. edn, 1966).

George, M. Dorothy, *Hogarth to Cruikshank: Social Change in Graphic Satire* (London: Allen Lane, 1967).

Godwin, William, *The Letters*, ed. Pamela Clemit, vols 1 and 2 (Oxford: Oxford University Press, 2014).

Gordon, Lyndall, *Mary Wollstonecraft: A New Genus* (London: Little, Brown and Co., 2005).

Gosse, Fayette, *The Gosses: An Anglo-Australian Family* (Canberra: Brian Clouston, 1981).

Gregory, Sir Richard, *British Scientists* (London: William Collins, 1941).

Grylls, Rosalie Glynn, *William Godwin and His World* (London: Odhams Press Ltd, 1953)

Hague, William, *William Pitt the Younger* (London: HarperPerennial, pb. edn, 2005)

Hale, J. R. (ed.), *The Italian Journal of Samuel Rogers* (London: Faber and Faber, 1956).

Haywood, Ian, and Seed, John (eds), *The Gordon Riots: Politics, Culture and Insurrection in Late Eighteenth-Century Britain* (Cambridge: Cambridge University Press, 2012).

Hazlitt, William, *The Spirit of the Age* (London: J. M. Dent and Sons Ltd, Everyman edition, 1910). Also Hazlitt, William, *The Spirit of the Age*, in *The Works of William Hazlitt*, The World's Classics (London: Grant Richards, 1904).

Henderson, Felicity (exhibition curator), *The Royal Society: 350 Years of Science*, catalogue for an exhibition in 2010 (London: The Royal Society, 2010).

Hibbert, Christopher, *George IV: Regent and King 1811–1830* (London: Allen Lane, 1973).

Hill, Draper, *Fashionable Contrasts: Caricatures by James Gillray* (London: Phaidon Press, 1966).

Hinde, Wendy, *George Canning* (London: Collins, 1973).

Holmes, Richard, *The Age of Wonder: How the Romantic Generation Discovered the Beauty and Terror of Science* (London: Harper Press, pb. edn, 2009).

Howell, Thomas Jones (compiler), *A Complete Collection of State Trials and Proceedings for High Treason and Other Crimes and Misdemeanors* [sic] *from the Year 1783 to the Present Time with Notes and Other Illustrations* (London:

T. C. Hansard, Peterborough Court, Fleet Street, 1817 and 1818), vols XXII, XXIV, XXV. In each volume there is a discrepancy between the number of the bound volume and the number on the title page, thus vol. XXII is vol. I on the title page, vol. XXIV is vol. III, vol. XXV is vol. IV. In each case the volume number given in the table of contents is the same as that of the bound volume.

Jewson, C. B., *The Jacobin City: A Portrait of Norwich in its Reaction to the French Revolution 1788–1802* (Glasgow: Blackie, 1975).

Johnson, L. G., *General T. Perronet Thompson 1783–1869, His Military, Literary and Political Campaigns* (London: George, Allen and Unwin Ltd, 1957).

Johnston, Kenneth R., *Unusual Suspects: Pitt's Reign of Alarm and the Lost Generation of the 1790s* (Oxford: Oxford University Press, 2013).

Keane, John, *Tom Paine: A Political Life* (London: Bloomsbury, 1995).

Kemble, Fanny, *Records of Later Life* (New York: H. Holt and Co., 1882).

Laboucheix, Henri, trans. Sylvia and David Raphael, *Richard Price as Moral Philosopher and Political Theorist* (Oxford: Voltaire Foundations at the Taylor Institute, 1982).

Lewin, C. G., *Pensions and Insurance before 1800: A Social History* (East Lothian: Tuckwell Press Ltd, 2003).

McCormmach, Russell, *Speculative Truth: Henry Cavendish, Natural Philosophy and the Rise of Modern Theoretical Science* (Oxford, New York: Oxford University Press, 2004).

Morgan, George Cadogan, *Lectures on Electricity* (Norwich: J. March, 1794, in Gale ECCO).

Morgan, George Cadogan, and Morgan, Richard Price, *Travels in Revolutionary France and A Journey Across America*, ed. Mary-Ann Constantine and Paul Frame (Cardiff: University of Wales Press, 2012).

Morus, Iwan Rhys, *Frankenstein's Children: Electricity, Exhibition, and Experiment in Early Nineteenth Century London* (Princeton: Princeton University Press, 1998).

O'Connell, Sheila, *London 1753*, catalogue for an exhibition at the British Museum in 2003 (London: The British Museum, 2003).

Ogborn, Maurice Edward, *Equitable Assurances: The Story of Life Assurance in the Experience of the Equitable Life Assurance Society 1762–1962* (London: George Allen and Unwin Ltd, 1962).

Owen, John B., *The Eighteenth Century 1714–1815* (New York: Norton 1974).

Paine, Thomas, *Rights of Man, Common Sense and Other Political Writings* (Oxford: Oxford University Press, pb. edn, 1995).

Peach, Bernard, 'On What Point did Richard Price Convince David Hume of a Mistake? With a Note by Henri Laboucheix', in *The Price–Priestley Newsletter*, 2 (1978), 76–81.

Peach, W. Bernard (ed.), *The Correspondence of Richard Price, vol. III: February 1786–February 1791* (Durham, NC: Duke University Press, 1994).

Picard, Lisa, *Dr. Johnson's London* (London: Phoenix, pb. edn, 2003).

Picker, John M., *Victorian Soundscapes* (Oxford: Oxford University Press, 2003).

Place, Francis, *The Autobiography of Francis Place (1771–1854)*, ed. Mary Thale (Cambridge: Cambridge University Press, 1972).

Porter, Roy, *English Society in the Eighteenth Century* (London: Penguin edn, 1982).

Price, Richard, 'An Essay towards Solving a Problem in the Doctrine of Chances, by the late Rev. Mr Bayes, F.R.S., communicated by Mr Price, in a letter to John Canton, A.M. F.R.S.', *Philosophical Transactions of the Royal Society*, 53 (1763), 370–418.

Price, Richard, 'Observations on the Expectations of Lives, The Increase of Mankind, The Influence of Great Towns on Population, and Particularly the State of London, with Respect to Healthfulness and Number of Inhabitants. Communicated to the Royal Society, April 27, 1769. In a Letter from Mr Richard Price F.R.S. to Benjamin Franklin, Esq.; LL.D. and F.R.S.', *Philosophical Transactions of the Royal Society*, 59 (1769).

Price, Richard, *Observations on Reversionary Payments; on Schemes for Providing Annuities for Widows, and for Persons in Old Age; on the Method of Calculating the Values of Assurances on Lives; and on the National Debt* (London: Cadell, 1771).

Price, Richard, *Political Writings*, ed. D. O. Thomas (Cambridge: Cambridge University Press, 1991).

Price, Richard and Horne Tooke, John, *Facts: Addressed to the Landholders, Stockholders, Merchants, Farmers, Manufacturers, Tradesmen, Proprietors of every Description, and generally all The Subjects of Great Britain and Ireland* (London, 1780).

Priestley, Joseph, *A Discourse on the Occasion of the Death of Dr Price delivered at Hackney on Sunday 1 May 1791* (London: J. Johnson, 1791).

Pullin V. E., and Wiltshire, W. J., *X-Rays Past and Present* (London: Ernest Benn, 1927).

Randall, H. J., *Bridgend: The Story of a Market Town* (Newport: R. H. Johns, 1955).

Ridley, Jane, *Bertie: A Life of Edward VII* (London: Vintage Books, pb. edn, 2012).

Rigby, Dr Edward, and Rigby Eastlake, Lady Elizabeth, *Dr Rigby's Letters from France &C. in 1789* (London: Longmans, Green, and Co., 1880, digitised by General Books LLC™, Memphis, TN, 2012).

Rocque, John (with introductory notes by Ralph Hyde), *The A to Z of Georgian London* London Topographical Society Publication No. 126 (Kent: Harry Margary, in association with Guildhall Library, London, 1982).

Rogers, Samuel, *Table-Talk & Recollections*, selected by Sir Christopher Ricks (London: Notting Hill Editions Ltd, 2011).

Ross, Cathy, and Clark, John, *London, the Illustrated History* (London: Penguin, 2011).

Schwitzer, Joan, *Buried in Hornsey: The Graves of St Mary's Churchyard* (London: Hornsey Historical Society).

Smollett, Tobias, *The Expedition of Humphry Clinker*, Oxford World's Classics edn (Oxford: Oxford University Press, 1966).

Sommers, Susan Mitchell, *The Siblys of London, A Family on the Esoteric Fringes of Georgian England* (Oxford: Oxford University Press, 2018).

Stephens, Alexander, *Memoirs of John Horne Tooke*, vol. II (London: J. Johnson and Co, 1813).

Strange, William, *Sketches of Her Majesty's Household interspersed with Historical Notes, Political Comments, and Critical Remarks* (London: William Strange, 1848).

Thomas, D. O., 'George Cadogan Morgan (1754–1798)', in *The Price–Priestley Newsletter*, 3 (1789).

Thomas, D. O., *The Honest Mind: The Thought and Work of Richard Price* (Oxford: Clarendon Press, 1977).

Thomas, D. O., and Peach, Bernard (eds), *The Correspondence of Richard Price*, vol. I, *July 1748–March 1778* (Cardiff: University of Wales Press, 1983).

Thomas, D. O. (ed.), *The Correspondence of Richard Price*, vol. II, *March 1778–February 1786* (Durham, NC: Duke University Press, 1991).

Thomas, J. E., *Britain's Last Invasion, Fishguard 1797* (Stroud: Tempus Publications Ltd, 2007).

Thomas, Roland, *Richard Price, Philosopher and Apostle of Liberty* (Oxford: Oxford University Press, 1924).

Thwaite, Ann, *Edmund Gosse: A Literary Landscape 1849–1928* (London: Secker and Warburg, 1984).

Thwaite, Ann, *Glimpses of the Wonderful: The Life of Philip Henry Gosse 1810–1888* (London: Faber and Faber, 2002).

Tucker, John V., 'Richard Price and the History of Science', in *Transactions of the Honourable Society of Cymmrodorion*, 23 (2017), 69–86.

Tucker, Tom, *Bolt of Fate: Benjamin Franklin and his Electric Kite Hoax* (Stroud: Sutton Publishing, 2004).

Turnbull, Craig, *A History of British Actuarial Thought* (London: Palgrave Macmillan, 2017).

Uglow, Jenny, *The Lunar Men: The Friends who Made the Future 1730–1810* (London: Faber and Faber, pb. edn, 2003).

Uglow, Jenny, *In These Times: Living in Britain Through Napoleon's Wars 1793–1815* (London: Faber and Faber, pb. edn, 2015.

Veitch, G. S., *The Genesis of Parliamentary Reform* (London: Constable, 1965).

Vincent, David (ed.), *Testaments of Radicalism: Memoirs of Working Class Politicians 1790–1885* (London: Europa Publications Ltd, 1977).

Wade, John, *History of the Middle and Working Classes; with a Popular Exposition of the Economical and Political Principles which have influenced the Past and Present Condition of the Industrious Orders* (London: Effingham Wilson, 1833).

Wade, John, *British History Chronologically Arranged; Comprehending a Classified Analysis of Events and Occurrences in Church and State; and of the Constitutional, Political, Commercial, Intellectual and Social Progress of the United Kingdom from the First Invasion by the Romans to AD 1847* (London: Henry G. Bohn, 1847).

Wharam, Alan, *The Treason Trials, 1794* (Leicester and London: Leicester University Press, 1992).

Waugh, Evelyn, *A Little Learning* (London: Chapman and Hall, 1964).

Waugh, Arthur, *One Man's Road* (London: Chapman and Hall, 1931).

Williams, Caroline E., *A Welsh Family From the Beginning of the Eighteenth Century* (London: Women's Printing Society Ltd, 1893, repr. Kessinger Publishing Legacy Reprints).

Wollstonecraft, Mary, *A Vindication of the Rights of Woman and A Vindication of the Rights of Men* (Oxford: Oxford World's Classics, pb. edn, 1999).

Unattributed, 'Account of the late Mr George Cadogan Morgan [Dec. 1798]', in *The Monthly Magazine and British Register*, vol. VI, no. XXXIX (London: printed for R. Phillips, December 1798).

Facsimile manuscripts

Macdonald, Angus S., et al., *Chance and Assurance: manuscripts of the work of Thomas Bayes, James Dodson, Richard Price and William Morgan; From the Archive of the Equitable Life Assurance Society preserved for research by the UK Actuarial Profession* (London: Institute of Actuaries, 2010) (DVD).

Websites and online books and articles

Balisciano, Dr Márcia, 'Benjamin Franklin and Public History', Introduction. Available at *http://www.benjaminfranklinhouse.org/site/sections/news/MBArticle.pdf*. Accessed 2015.

Gurney, Joseph, *The Trial of Thomas Hardy for High Treason, at the Sessions House in the Old Bailey* (London: Martha Gurney, Bookseller, Holborn-Hill, 1795). Available at *https://books.google.co.uk/books?id=HHc2AAAAMAAJ&pg=PA229&dq=the+trial+of+john+horne+tooke&hl=en&sa=X&ved=0ahUKEwjjw5Skrffo AhWZFcAKHUrpB9wQ6AEIQTAD#v=onepage&q=the%20trial%20of%20 john%20horne%20tooke&f=false*. Accessed 2019.

House of Commons, *Report from the Select Committee on the Poor Laws: with the minutes of evidence taken before the committee: and an appendix* (London, July 1817). Available at *https://babel.hathitrust.org/cgi/pt?id=nyp.33433000256689 &view=1up&seq=142*. Accessed 2018.

INDEX